CLEAR AND CONCISE

CLEAR AND CONCISE

Technical Writing for Biological Engineers

ABIGAIL S. ENGELBERTH

PURDUE UNIVERSITY PRESS

WEST LAFAYETTE, INDIANA

Cataloging-in-Publication Data on file at the Library Congress.

978-1-62671-319-2 (paperback)
978-1-62671-320-8 (epdf)

Cover images: Vector unreadable handwriting, crossed out phrases: Larisa Zaytseva/iStock via Getty Images Plus. Science icons set: PeterSnow/iStock via Getty Images Plus.

*To T. C. Brown, loved from the very
beginning until the never-ending.
For my children—may they receive clear guidance and
superb coaching wherever their paths may lead.*

CONTENTS

PREFACE

A Journey from Classroom to Publication

This book has been a long labor of love. It began in 2014 as a simple "Guide for Success" for students in ABE 304: Bioprocess Engineering Laboratory at Purdue University. When Dr. Jenna Rickus and I first taught ABE 304 in 2013, we quickly realized that students needed more than lab instructions—they needed clear, consistent guidance on how to write technical reports. Without this guidance, even strong students submitted work that lacked structure and clarity. My goal was to give students as much context as possible so they could build a solid foundation in technical writing.

The first guide was just sixteen pages long, drawing heavily from my own undergraduate lab experience and early teaching notes. Over time that guide grew—little by little—into something much more comprehensive. Each semester, I added new sections, examples, and checklists based on what students struggled with most. Student feedback played a huge role in shaping this book. They consistently asked for examples, and while I sometimes hesitated to provide a single "perfect" model (because there is no one right way to write well), I learned that showing real examples—paired with revisions—helped students see that good writing is a process of improvement.

Turning years of course materials into a publishable book was both exciting and intimidating. The biggest challenge was tone. Some sections I wrote years ago were formal and rigid, while newer material reflected a more coaching-oriented style. That evolution mirrors my own growth as an instructor. I've learned that students respond best when the guidance feels practical and encouraging, not just prescriptive.

This book is a testament to continuous improvement—both mine and that of my students. It blends classroom-tested strategies, anecdotal lessons, and best practices from trusted resources. Alumni often tell me how much the original guide helped them long after graduation, and that feedback gave me the confidence to share this work more broadly.

Finally, I do want to acknowledge that I used Copilot during the preparation of this book. It helped me catch inconsistencies in tone and brainstorm ways to expand the original course materials into full chapters. While every decision and every word ultimately reflects my judgment and experience, Copilot served as a valuable tool for refining ideas and ensuring clarity.

My hope is that this book will do the same for you: provide clarity, build confidence, and help you see technical writing not as a hurdle but as a skill that will serve you throughout your career.

1

WHY LEARN TECHNICAL WRITING

Understanding the Role of Communication in Engineering Practice

OBJECTIVES

- Understand the purpose and scope of technical writing in engineering contexts.
- Distinguish technical writing from other forms of writing.
- Recognize the key competencies required for effective technical communication.
- Describe the stages of the technical writing process.
- Apply foundational principles to evaluate and improve technical writing.

WHAT IS TECHNICAL WRITING?

Technical writing is all about transferring knowledge from one person to another. It is not just about words; it is about making ideas clear, accessible, and useful. That is why we use more than just text. Diagrams, charts, and other visual aids are powerful tools that help convey complex information quickly and effectively.

When used thoughtfully, visuals can elevate your writing. A well-placed figure with clear labeling and explanation can make your work easier to understand and more impactful. Visuals should serve a purpose—they are not a mere decoration. They should help your reader grasp the structure of a system, see trends in data, or understand how a process is connected.

Think of your writing as a conversation. Your readers may not have the same background or context you do, so your job is to guide them. Visuals are part of that guidance. When you include a chart, a schematic, or a diagram, you are giving your readers another way to access the information—a way to really *see* the concepts.

As you revise your work, ask yourself these questions: Does this visual help my reader understand the point I'm making? Is it labeled clearly? Could someone understand it without reading the entire text? If the answer is yes, you are on the right track.

A NATURAL WRITER?

Many students assume that good technical writing is a natural talent, but this is rarely the case. Writing clearly and concisely about complex technical topics is a learned skill, developed through practice and feedback. Being strong in general English or creative writing does not automatically translate to success in technical writing because the goals are different: technical writing prioritizes clarity, precision, and reproducibility over style and personal expression.

Like any engineering skill, writing improves with deliberate effort. Each report you write is an opportunity to refine your ability to organize information, explain processes, and present data objectively. Do not be discouraged if your first drafts feel awkward—most professionals spend more time revising than writing. Treat writing as part of your engineering tool kit. The more you practice, the more confident and effective you will become.

HOW IS TECHNICAL WRITING DIFFERENT?

Technical writing should be precise and unambiguous. Avoid personal voice and stylistic flourishes; they distract from clarity and can introduce bias. Do not use persuasive, hyperbolic, or metaphorical language—these belong in creative or literary writing, not in technical documents. Your goal is to communicate facts and processes so that any technically competent reader can understand and reproduce the work without guessing your intent. Every sentence should serve a purpose: to inform, explain, or document.

It is often obvious when a writer is unhappy with their results, but technical writing should never reveal personal feelings about the outcome. Do not express frustration, disappointment, or excitement in your report. Instead, present the information objectively, explain what the data means, and move on. Your role is to analyze and interpret, not to editorialize. When in doubt, choose clarity over creativity and accuracy over emotion.

TECHNICAL WRITING PROCESS

To be a proficient technical writer, you should follow a structured process (Figure 1.1) to ensure that complex information is communicated clearly and effectively. The process typically involves the following key steps:

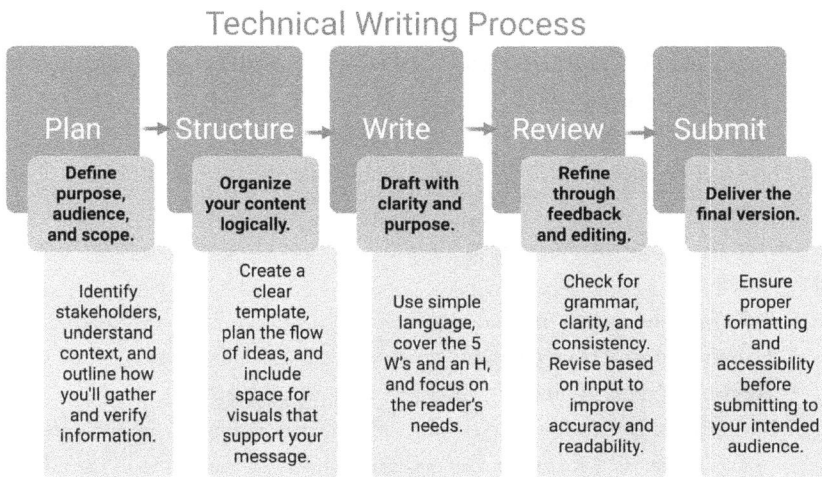

FIGURE 1.1 Overview of the technical writing process. Created by the author in BioRender (https://BioRender.com)

1. **Plan**

   Before writing begins, it is essential to define the **scope**, **purpose**, and **audience** of your document. This stage includes
   - Identifying key **stakeholders** and their expectations,
   - Understanding the **context** in which the document will be used, and

- Outlining the **process** for gathering and verifying information.

A well thought-out plan sets the foundation for a focused and relevant document.

2. Structure

Once the plan is in place, the next step is to organize your content:

- Develop a **template** with logical **headings and subheadings**.
- Consider the **flow of information**—from general to specific or problem to solution.
- Incorporate placeholders for **visual aids** such as charts, diagrams, and tables that support the text.

A clear structure helps your reader navigate the document and locate key information quickly.

3. Write

Now you can begin drafting your content:

- Use the **KISS principle** (Keep it simple, stupid) to maintain clarity and avoid jargon.
- Apply the **5 W's and an H** (who, what, when, where, why, and how) to ensure completeness.
- Write in a **concise, objective, and reader-focused** style.

This stage is about translating technical knowledge into accessible language.

4. Review

After drafting, the document should go through a thorough review process:

- **Edit** for grammar, clarity, and consistency.
- **Solicit feedback** from peers, subject matter experts, or intended users.
- **Revise** based on comments to improve accuracy and readability.

Multiple rounds of review help ensure that the document meets its goals and is free of errors. Most writers spend the majority of their time in the editing phase. Do not be discouraged if you find yourself spending much of your time editing your work.

5. Submit

The final step is to deliver the document to its intended audience:

- Submit the document to an **instructor**, a **supervisor**, or a **client**.
- Or publish it for **wider dissemination**, such as in a report repository, a website, or a journal.

Ensure that the final version is formatted correctly and is accessible to all intended readers.

PRINCIPLES TO GUIDE YOUR WRITING

Technical writing should be clear, concise, and purposeful. Brevity, simplicity, clarity, precision, and plainness are among the key principles to follow when writing.

Brevity is crucial in technical writing. Concise language ensures that the message is clear and to the point. Use as few words as necessary to convey the meaning. Example:

> ✗ *Due to the fact that the experiment was unsuccessful, we decided to repeat it.*
> ✓ *The experiment was repeated due to initial failure.*

The revised sentence eliminates unnecessary words, making the message more direct and efficient.

Simplicity ensures that complex information is easy to understand. Avoid jargon and overly complex language unless absolutely necessary. Example:

> ✗ *Utilize the apparatus to initiate the combustion sequence.*
> ✓ *Depress the ignition button to start the fire.*

The revised sentence replaces jargon with familiar terms, making the instruction easier to understand.

Clarity ensures that your reader understands exactly what you intend to convey. Avoid ambiguity by choosing precise words and constructing logically ordered sentences. Each sentence should have a clear subject and action, and transitions between ideas should be smooth and purposeful. Example:

> ✗ *We tested the samples after storing them in the freezer and incubator.*
> ✓ *We tested the samples after storing them—first in the freezer, then in the incubator.*

The revised sentence uses a clear sequence and precise wording to avoid ambiguity.

Choosing strong, active verbs enhances **precision** in technical writing. Precise verbs eliminate vagueness and clarify the action being described, making the writing more direct and informative. Passive constructions and weak verbs can obscure meaning or reduce impact.

Example:

✗ *The solution was subjected to heating.*
✓ *We heated the solution.*

The revised sentence clearly identifies the actor and the action, improving both clarity and efficiency.

Plainness helps maintain clarity and focus in writing. Avoid convoluted phrasing when simpler language will suffice.

Example:

✗ *The results may be indicative of a potential correlation between the variables.*
✓ *The results suggest a possible link between the variables.*

The revised sentence uses familiar, straightforward language that improves accessibility and eliminates unnecessary complexity.

2
TECHNICAL WRITING TIPS AND GUIDELINES

Principles, Style, and Strategies for Clear Engineering Communication

---

OBJECTIVES

- Apply principles of clarity, conciseness, and precision in technical writing.
- Identify and avoid common stylistic and formatting errors.
- Write for a technically literate but unfamiliar audience.
- Use appropriate tone, tense, and structure in engineering documents.
- Recognize the role of formatting and presentation in professional communication.

GENERAL WRITING PRINCIPLES

Be concise in your writing. Do not use three sentences when two will do. Avoid being overly verbose. Being concise in your writing ensures that your message is delivered efficiently, without distracting or overwhelming the reader. In technical contexts, readers often seek specific information quickly—excessive wording can obscure key points, slow comprehension, and reduce the overall impact of your communication. By saying what needs to be said and no more, you respect your reader's time and improve the clarity and precision of your work.

Write with the attitude that what you are writing will be for your job, as though your pay or promotion depends on your communication. In your career, your writing will often be the only thing people see before they make decisions about your work, your competence, and even your potential. If you treat every report like it could influence your job, your pay, or your promotion, you'll naturally write with more care, precision, and professionalism. This mindset helps you build habits that will serve you well beyond the classroom.

A considerable amount of judgment must be exercised when you write: the ideas to be discussed and how best to organize these, the detail to be provided, the figures and tables to be shared, the comparisons with theory or other work, etc. Good writing takes judgment and time. You must decide what ideas are worth discussing, how to organize them, how much detail to include, and which figures or comparisons will help your reader understand your work. These decisions are not automatic—they take time, and they require you to think carefully about your audience, your purpose, and your message. The more intentional you are in making these choices, the stronger and more effective your writing will be.

Make your work look good. Be neat and well organized. Pay attention to formatting. Use headings and subheadings to let readers know what they will be reading. Presentation matters. If your writing is messy, disorganized, or difficult to follow, your readers will assume that your thinking is the same. Formatting is more than just aesthetics; it guides your readers through your ideas. Headings and subheadings help your audience to know what to expect from each section and where to find key information. Neat, well-organized work shows that you care about your message and respect your readers' time.

AUDIENCE AND TONE

Write with a level of complexity that is consistent with your technical background. You have acquired a lot of knowledge in your studies. Show your readers what you know. Your writing should reflect the depth of your understanding. This doesn't mean using jargon for the sake of sounding smart, but it does mean writing with the confidence and precision of someone who knows the material. If your writing is too basic, it undersells your expertise. If it's too complex, it risks losing your readers. Find the balance that shows that you know your stuff and can communicate it effectively.

Write for an audience that also has a technical background but is most likely not entirely familiar with your experiment. Assume that your readers know the field but not your specific work. This means that you do not need to explain basic concepts such as how a spectrophotometer operates, but you do need to explain how you used it and why. This kind of audience-aware writing helps you avoid both underexplaining and overexplaining. It also forces you to think critically about what is essential for understanding your experiment and what can be assumed.

Write in third person and use past tense. You are explaining what you have done, not what you intend to do. Technical reports are records of completed work, not plans of what you will do or personal reflections. Writing in third person and past tense keeps the focus on the experiment and the results. It also aligns with professional and academic norms, making your writing feel more formal and objective. "We poured the solution" becomes "The solution was poured," which shifts attention to the process and outcome, where it belongs.

Avoid using contractions. Contractions make writing feel casual, and casual is not what you're going for in a technical report. You want your writing to sound polished and professional. "Don't" becomes "do not," and "can't" becomes "cannot," while "ain't" is not a word—or at least that is what my Mother told me. Small changes in word choice help your tone stay consistent with the expectations of scientific and engineering communication.

CONTENT AND STRUCTURE

Avoid filler phrases such as "The purpose of this lab was . . ." and "For this lab we . . ." Filler phrases such as "The purpose of this lab was . . ." do not add value and often signal that you are writing to fill space rather than communicate something meaningful. Instead, get straight to the point.

Example:

- ✘ *The purpose of this lab was to determine the viscosity of the fluid using a rotational viscometer.*
- ✓ *Viscosity was determined using a rotational viscometer.*

Do not use titles such as "Rheology Lab." Be descriptive, but keep it concise. Titles such as "Rheology Lab" are too vague to be useful. A good title should give your reader a sense of what the report is about without being wordy. Try to convey the purpose of the report through the title.

TERMINOLOGY AND DEFINITIONS

All variables and uncommon terms must be defined. All acronyms need to be spelled out at least once at their first mention. Do not make your reader guess the meaning of a particular term. If you introduce a variable, define it clearly so your reader knows exactly what it represents. Spell out acronyms the first time to avoid confusion. Even if the term feels familiar to you, it might not be to someone else reading your report. It may also be that an acronym has a different meaning based on the background of the reader. In the Department of Agricultural and Biological Engineering, "BMP" can mean either "best management practices" or "bone morphologic proteins," which are very different things. Defining terms up front shows respect for your audience and helps ensure that your writing is accessible and accurate. Avoid using an acronym simply because it exists. Acronyms should only be used if the term is mentioned three or more times in your lab reports; other types of reports or publications may have different conventions Once you have spelled out the acronym, use only the acronym from there on out and not the spelled-out term.

A special note regarding scientific names. When reporting biological organisms in your writing, always spell out the full scientific name the first time it appears. After that, you may abbreviate by using the first letter of the genus followed by the full species name.

> ✓ *Example: Saccharomyces cerevisiae* is the most common microorganism used in the fermentation of beer. Top fermenting strains of *S. cerevisiae* are used to make ales, while bottom fermenting strains are used to produce lagers.

Common terms such as "psi" and "mL" do not need to be defined. Do not overexplain. Terms such as "psi," "mL," and "°C" are standard in technical writing and widely understood by your audience. Defining them wastes space and can make your writing feel clunky. Use your judgment. If a term is commonly used in engineering or science, you can assume that your reader knows it. Save your explanations for the things that are specific to your experiment or less familiar to your audience.

Pay attention to pluralization when discussing dimensionless numbers. For example, "Reynolds number: should be written as shown here, not as "Reynold's number." The term is not possessive. The ratio is named after Osborne Reynolds, whose last name is Reynolds. In fact, none of the dimensionless numbers are possessive because they represent changing ratios. In contrast, Avogadro's number is possessive, as it refers to a fixed value—literally his number.

VISUALS AND DATA DISPLAY

Prepare tables, graphs, and figures very carefully. Visuals are not just a pretty picture to add to your report; they are a crucial part of technical communication, as discussed in Chapter 1. A sloppy figure or a poorly constructed table can confuse your reader or even misrepresent your data. Take the time to make sure your visuals are accurate, clean, and easy to interpret. Something that looks rushed or unclear reflects poorly on the quality of your work. Your figures should help your readers understand your results, not make them work harder to figure out what you are trying to convey.

Label all visuals appropriately. Every table, graph, and figure should include clear labels: axes, regression lines, components of an apparatus, and anything else your reader needs to interpret the visual accurately. Do not make your reader guess. Labels provide context and meaning by showing what is being presented and how it connects to your analysis. Without clear labels, even strong data can lose their impact. Think of labels as part of your explanation; they guide your reader through the story your data tells.

See Chapter 6 for additional guidance on visuals and data display.

REFERENCES AND ATTRIBUTION

Use your own words to avoid plagiarism. It is not difficult to recognize when material has been copied, and plagiarism will be heavily penalized. Do not use direct quotes from your sources; synthesize what you read and put it into your own words. Copying someone else's words—even when you include a citation—is not the same as demonstrating understanding. When you write in your own words, you show that you have processed the information and can communicate it clearly. Reports are how we assess your understanding of the information. Plagiarism is often easy to spot and undermines your credibility. Direct quotes do not belong in technical writing; your report is not a comparative literature paper, and you are too far along in your education to not be able to synthesize information. Instead, read your sources, think about what they mean, and explain the ideas in a way that fits your report.

Use references as needed. If you state a fact from a source, you need to cite the source. If you did not generate the information yourself, you need to tell your reader from where it came. That is how we maintain transparency and academic integrity. Facts in your reports generally come from either your experimental findings or external sources used for comparison. Citing those sources shows that you have performed proper research and allows your readers to follow up if they want to learn more. Citing is not optional. If you use a fact, a method, or a comparison from a source, you must cite it.

Cite references appropriately. Look in a textbook or journal article for style examples. Sloppy or inconsistent citations make your work look careless and can confuse your reader. Use a consistent format that aligns with professional standards. Every sentence that refers to an external source must include proper attribution. Your reader should always know where the information came from and how it supports your analysis.

See Chapter 9 for additional guidance on references and citations.

FORMATTING AND STYLE

Do not begin a sentence with a number (i.e., 10 mL of water was poured into a beaker). Starting a sentence with a numeral is visually jarring and grammatically incorrect in formal writing and disrupts the flow and can confuse your reader.

Numerals less than 10 should be spelled out unless it is a measured or quantitative value. Spelling out numbers under 10 helps maintain consistency and readability. A measured values is reported as a numeral because it is clearer and more appropriate. This rule helps you strike a balance between professionalism and clarity.

✓ *Example:* Five pressures were measured ranging from 10 psi to 25 psi.

Include the leading zero in a measured value (i.e., 0.75 mL, not .75 mL). Leaving off the leading zero makes the number harder to read and easier to misinterpret, especially in technical writing where precision matters. A value such as .75 can be missed or misread in a dense paragraph or data table. Including the zero improves readability and ensures that your data is communicated clearly and professionally.

Use superscripts and subscripts as required. It is the 21st century, and you have access to sophisticated word processing software, not a typewriter, and do not need to rely on carets to indicate baseline.

Use proper scientific notation. "1E4" is incorrect notation. "1×10^4" is the correct format; it is clear, consistent, and immediately recognizable to your reader. Using proper notation shows that you understand the standards of technical communication and care about presenting your work accurately. Yes, Microsoft

Excel and Google Sheets use E as a shorthand, but this type of shorthand is not appropriate for a report.

REPORT CONTENTS

A well-structured report helps your reader follow your thinking and better understand your work. Each section has a distinct purpose, and skipping or blending sections makes your writing harder to follow. Think of your report like a story: it needs context, action, and resolution.

- **Beginning**
  - Goals and introduction
    - This sets the stage. What were you trying to accomplish? Why does it matter? Your reader needs to know the purpose before diving into the details.
  - Background
    - Provide context. What's already known about this topic? What led you to this experiment? This helps your reader understand how your work fits into the bigger picture.
  - Basic Principles
    - Explain the science or engineering concepts that underpin your experiment. This shows that you understand the theory and helps your reader follow your analysis later.
  - Related Models and Theories
    - Mention any existing models or theories that your work may build upon. This gives your reader a framework for interpreting your results and understanding your comparisons.
- **Middle**
  - Nature of Experiment
    - Describe what you did. What was the experiment designed to test or measure? This section should be clear and focused.
  - Equipment Used
    - List and describe the tools and instruments you used. Your reader should be able to understand how the data was collected and whether the equipment was appropriate. Often a visual of a particular piece of equipment is helpful to include to provide additional context.
  - Procedures

- A concise, well-organized methods section gives your reader a clear understanding of the experimental design without becoming bogged down in unnecessary detail. Focus on what matters: what was performed, what was measured, and in what sequence. Highlight the decisions and conditions that shaped the experiment.
  - Data to Be Obtained
    - Clarify what kind of data you were expecting to collect and why. This helps your reader anticipate the analysis and understand your goals.
- **End**
  - Data Analysis
    - Show how you processed and interpreted the data. This is where you demonstrate your ability to connect raw results to meaningful conclusions.
  - Results Obtained
    - Present your findings clearly. Use visuals where appropriate, and make sure your reader can see the connection between your procedures and your outcomes.
  - Error Analysis
    - Briefly acknowledge limitations. What could have affected your results? What sources of error were present, and how significant were they? This shows maturity in your scientific thinking.
  - Conclusion
    - Wrap it up. What did you learn? Did the experiment meet its goals? What are the implications of your findings? This is your chance to reflect and tie everything together.

REPORT FORMATTING

All reports should be written in Times New Roman 11-point font with one-inch margins on all sides. The body text should be left-justified and use 1.5 line spacing to ensure readability. You may use single spacing for table and figure captions to visually separate them from the main text.

If an instructor, supervisor, or agency provides different formatting guidelines, follow those instead—these guidelines are often given for a specific reason. For example, the line-spacing recommendation here is based on readability: single-spaced text is difficult to grade on a screen, while double spacing tends to feel too spread out. Formatting may seem like a small detail, but it plays a big role in how your work is received. Clear formatting helps your reader focus on your ideas without distraction.

REPORT FINAL CHECK

After writing your report, ask yourself the following questions to ensure that it meets professional and academic expectations:

1. Does the report have something meaningful to say?
2. Is the report in a format acceptable to the reader?
3. Are the ideas organized in a way that makes it easy for your reader to understand?
4. Have you used an appropriate level of sophistication in your writing?
5. Does the report look good?
6. Have you made effective use of figures and tables?

COLLABORATIVE WRITING AND PEER REVIEW

Scientific and technical writing is often a team effort. Whether you are working on a lab report, research paper or a proposal, collaboration can improve the clarity, accuracy, and overall quality of your work if done effectively.

STRATEGIES FOR WRITING IN TEAMS

Collaborative writing requires coordination and clear communication. Here are some strategies to help:

- Divide the work strategically: Assign sections based on team members' strengths or interests (e.g., one person writes the methods, and another person writes the results).
- Maintain consistency: Agree on formatting, tone, and citation style early to ensure a unified final product. Assign one person to read through and edit the report before submission to help ensure that it reads as though it is from a single author.
- Use a shared outline: Start with a common structure so that everyone understands how their contributions fit into the whole.

- Designate a final editor: One person should be responsible for integrating and polishing the final draft to ensure coherence.

TOOLS FOR COLLABORATION

While platforms such as Google Docs and Sheets are popular, Microsoft Teams and the Office suite (Word, Excel, PowerPoint) offer greater functionality and integration, especially for technical and data-driven projects.

- Microsoft Word: Supports advanced formatting, citation tools, and real-time coauthoring. Ideal for structured documents with embedded figures, tables, and references.
- Microsoft Excel: Offers powerful data analysis, charting, and formula capabilities that far exceed those of Google Sheets.
- Microsoft Teams: Provides a centralized space for communication, file sharing, and task coordination. Teams integrate seamlessly with Word and Excel, making it easy to collaborate on documents while discussing edits or assigning tasks.

These tools are especially useful in academic and professional settings, where precision, formatting control, and version history are critical.

GIVING AND RECEIVING CONSTRUCTIVE FEEDBACK

Peer review is more than just editing; it is a process that builds critical thinking and reflection skills. Reviewing others' work helps you practice the cognitive skills needed to improve your own writing (Andersson & Weurlander, 2019). The following guidelines are provided to help make peer review productive:

- Focus on clarity, logic, and evidence. →Suggest concrete improvements.
- Frame feedback as suggestions, not criticisms. →Use constructive language.
- When receiving feedback, listen carefully and ask clarifying questions. →Treat feedback as an opportunity to improve.

BENEFITS OF PEER REVIEW

Peer review helps in the following areas:

- Improves clarity: Others can spot confusing phrasing or gaps in logic.
- Catches errors: A second set of eyes can identify mistakes in calculations, citations, and formatting.
- Strengthens arguments: Feedback can help refine reasoning and better support conclusions.
- Builds confidence: Knowing that your work has been reviewed and improved by others can make you more confident in submitting or presenting it.

ARTIFICIAL INTELLIGENCE FOR REVIEW

Artificial intelligence (AI) is rapidly transforming how we write, review, and revise technical documents. In Purdue's ABE 304: Biological Engineering Laboratory class, students use Charlie, Purdue's AI writing assistant integrated into the Circuit peer review platform, to receive feedback on drafts of their lab reports. Charlie provides suggestions on clarity, structure, and technical writing conventions, helping students refine their work before submission.

AI tools such as Charlie are meant to augment human judgment, not replace it. By identifying areas for improvement and offering real-time suggestions, AI can help students become more confident and independent writers.

AI tools are evolving quickly. New features and platforms are released regularly, and their capabilities are expanding. Stay curious but always verify AI-generated suggestions and maintain academic integrity.

REFERENCE USED IN THIS CHAPTER

Andersson, M., & Weurlander, M. (2019). Peer review of laboratory reports for engineering students. *European Journal of Engineering Education, 44*(3), 417–428.

3
WRITING A
FORMAL-STYLE LAB
REPORT

Structuring and Presenting Experimental Work with Professional Standards

---

OBJECTIVES

- Structure a formal lab report using standardized section headings.
- Develop clear and informative summaries, introductions, and conclusions.
- Present experimental methods and results effectively.
- Integrate theory and analysis to support findings.
- Use standardized formatting and headings to meet professional expectations.

FORMAL REPORT SECTIONS

Formal reports are divided into sections, which have the following distinctive headings:

1. Title Page   (5 points)
2. Summary   (10 points)
3. Introduction   (5 points)
4. Theory and Basic Principles   (10 points)
5. Methods   (10 points)
6. Results and Discussion   (20 points)

7. Conclusions and Recommendations   (10 points)
8. Nomenclature   (2.5 points)
9. Literature Cited   (2.5 points)
10. Appendices   (5 points)
Overall format of the report   (20 points)

Sections 2, 3, and 8 should each begin on a new page. Use the exact headings provided for your report to help guide your audience and make it clear which section they are reading. Consistent headings are more than simple organization; they reflect your ability to follow instructions and meet expectations. Many journals, companies, and professional organizations have strict formatting guidelines, and learning to follow them now builds habits that will serve you well later. When you use the correct headings, you make your report easier to navigate and demonstrate that you understand how to communicate within a technical framework. The following outline offers guidance on the typical content of each section. These suggestions are not absolute, but they reflect common practice and expectations in technical writing.

TITLE PAGE

The title page should clearly convey the purpose of the report and identify the individuals who completed the work. Use a specific and descriptive title. For example, "Effect of Pump Configuration on Flow of Ethanol through a Piping System" is more informative than a generic title such as "Pumps and Piping Lab Report." Be sure to include the submission date and write out your full names to take proper credit for your work.

SUMMARY

The summary is intended to completely yet briefly inform the reader of the basic nature and major implications as a **one-page** synopsis on the major aspects of the work. This section should state the scope of the work, briefly what was done and how, the principal results, the major conclusions that were drawn, and the significance of the work. The summary should be able to *stand alone*. References to other parts of the report should not be used in the summary. In a professional setting, the summary would be used to determine if the report was worth reading in its entirety. You can think of the summary as an abstract

INTRODUCTION

The summary is intended to stand alone and should be written as if the main body of the report does not exist. Likewise, the main body should include an introduction that assumes that the summary is not present. The introduction should briefly outline the scope and relevance of the investigation, describe the general techniques used, and clearly state the objective of the work. Be explicit about what you aimed to discover through the experiment. The introduction should align with the conclusion, so keep that connection in mind while writing. Do not include any results or conclusions in the introduction, as those will be addressed in later sections.

THEORY AND BASIC PRINCIPLES

A discussion of the theory or basic principles involved with this project is appropriate to include. Assume that your reader is technically proficient but not necessarily acquainted with the specific background of your work. This section should provide enough detail so that the reader will understand the basis for the laboratory work and the analysis of the data. For example, your reader will know what a heat transfer coefficient is, but you will need to explain how it relates to your experiment. Use equations when needed, and define the symbols. Use references for your equations and state conditions at which they are valid. In this section, your aim is to present what will be studied and the necessary foundations needed to understand what you will present in the *Results and Discussion*. Your *Theory and Basic Principles* section should mirror the order of the *Results and Discussion* section. You are guiding your reader through the analysis.

METHODS

Use this section to explain the experimental equipment and procedures. Explain major items using computer-drawn renderings of the equipment or process to help the reader visualize the experiment. **If a sketch is required, it must be drawn using software on a computer.** Oftentimes flow diagrams and line drawings are sufficient.

Do not get too bogged down in exhaustive detail for familiar procedures. For example, indicating that "product concentration was determined by titrating samples from the reactor with standard acid" is a sufficient description. You do not need to say that a 10 mL sample was taken and placed in a 250 mL Erlenmeyer flask and then titrated with 0.05N HCl using Bromothymol blue as the indicator. However, be sensitive to the fact that in certain instances procedural details form a key part of the information to be conveyed in the report. In most instances, provide a general overview of what was performed, what was measured, and in what sequence. Highlight the important procedural items.

Aspects such as how many trials were conducted, how conditions were varied, what ranges of variables were investigated, etc. are of interest in this section. Do not write as a step-by-step account. This is a methods section, not a procedure. Chapter 7 goes into detail for how to convert a procedure into a method and gives an example of a well-written methods section.

RESULTS AND DISCUSSION

This section is the heart of the report—the place where you demonstrate the value of your work and your understanding of its implications. *Results* refers to the presentation of your findings in text, graphs, tables, or other formats. *Discussion* involves interpreting those findings and assessing their significance. Choose the format that best communicates your results clearly and efficiently. Graphs are often preferred because they make trends easier to identify than is the case with tables. Refer to textbooks, journal articles, or other scientific writing for examples of effective presentation.

When introducing a result, briefly explain how the data were processed or analyzed and then present the corresponding figure or table. For example:

Figure 1 shows the effect of carrier gas flow rate on the retention times for benzene and acetone.

Follow this with a discussion of the figure's key features. Are the trends expected? Do the results align with theory or previous studies? *After* sufficient elaboration, move on to the next figure or table.

Figures should be combined when doing so improves clarity or facilitates comparison. While saving space is one reason to combine figures, the more important reason is to help the reader easily compare related data. A useful rule of thumb is that if your figure captions differ only by a descriptor (e.g., trial number or condition), consider combining the data into a single figure with clear labeling. This approach reduces redundancy and enhances interpretability.

Use appropriate significant figures when presenting numerical results.

At the end of this section, it is often helpful to provide a collective discussion of all results, summarizing key patterns and insights.

When interpreting your results, be realistic. Do not force your data to fit a theory. Critically examine sources of experimental error and discuss them *if* they significantly affect your findings. Consider whether the assumptions behind the theory were met in your experiment. If not, assess how this might impact the agreement between theory and results. If your data show significant scatter and no clear correlation, report this honestly and explore possible reasons for the variability.

Avoid drawing conclusions based on preconceived expectations. However, if your data are conclusive, reveal important phenomena, or show strong agreement with theory, be sure to highlight those successes.

CONCLUSIONS AND RECOMMENDATIONS

Major conclusions drawn from the experiment can be listed (1, 2, 3, etc.) *following* a simple introductory statement. Alternately, a conventional narrative style may be used. A well-written conclusion will tie in with the introduction. Use the conclusion to answer the objective posed in the Introduction. A list of recommendations for experimental alternatives can be included.

An example cover page for a formal report is found on the following page.

Final Report

Effect of Pump Configuration on Flow of Ethanol through a
Piping System

A. B. Estudent

Course Name
Agricultural & Biological Engineering
Purdue University
West Lafayette, IN 47907

Month Day, Year

4
MEMO-STYLE REPORTS

Communicating Results Quickly and Effectively for Decision-Makers

---

OBJECTIVES

- Understand the purpose and format of memo-style technical reports.
- Communicate key findings and recommendations concisely.
- Incorporate visuals and data into a narrative format.
- Tailor writing for quick comprehension by decision-makers.

STRUCTURE OF A MEMO REPORT

A memo report is intended to communicate fully the results of the work completed without providing the exhaustive detail that normally accompanies a more formal report. The physical style is that of a long memorandum or business letter and makes use of a continuous narrative rather than sections with formal headings. The report begins with a single page that conveys the key concepts and takeaway. The body of the report following the cover page can then runs two to three pages in length, with key tables, figures in the body, and data and sample calculations included in the appendix. Your report should include up to—and no more than—three key figures in the body of the report; other pertinent supplemental figures can be placed in the appendix. An example cover page and a body opening are included. The following sequence of development is suitable:

1. Cover page (i.e., PowerPoint slide) with objective, method, expectation, and takeaway message (20 points). This will be most beneficial to your

project manager, who can quickly discern the results and then read through the report for further clarification.

2. Brief introduction and orientation to the work carried out (10 points), a concise description of the scope of the work, and a summary of the relevant background *if needed*. The relevant theory and references to other work should be **integrated** within the presentation of the results.

3. A brief description of the experimental apparatus and methods used (10 points). Use a computer-drawn sketch to help explain. If needed, the sketch can be placed in the appendix. Note that not all experiments will need a sketch, as they use only a single piece of standard equipment.

4. Present and discuss results (20 points). Use figures and tables as appropriate. Provide comparisons to theory or experimental results of others. Assess the significance of the results. Do they agree with the literature sources? What sources of error exist?

5. Complete the text by briefly summarizing the major conclusions (10 points).

6. Append (10 points) a list of citations, nomenclature, sample calculations, and details necessary to supplement the text.

Overall format of the report (20 points)

WHY A MEMO?

A memo report is useful when your supervisor needs a quick, high-level understanding of your work (i.e., before a meeting or conference). This format is particularly effective when information needs to be communicated concisely and efficiently. A well-written memo report can also support continued funding for your project by clearly demonstrating progress and value in a compact form.

An example cover sheet is provided in Figure 4.1. The cover page will fill one entire sheet when printed in landscape setting.

Membrane Control Regime

Experiment Performed by: A.B. Estudent

Objective: Determine the flux regime of an ultrafiltration membrane by varying transmembrane pressure and pump flow rate

Method: A 500 kDa ultrafiltration membrane was used to concentrate a 0.25% (w/v) solution of pure xanthan gum. Flow rate [Q] and Pressure [ΔP] were varied and resulted in a resulting permeate flux was calculated.

Expectation: A static response would indicate a mass-transfer controlled membrane while a positive linear relationship would indicate that the membrane was controlled by the change in pressure.

Take-a-way: The membrane is mass-transfer controlled as indicated by the nearly unchanged relationship between ΔP and Permeate Flux.

Expected Result

Actual results

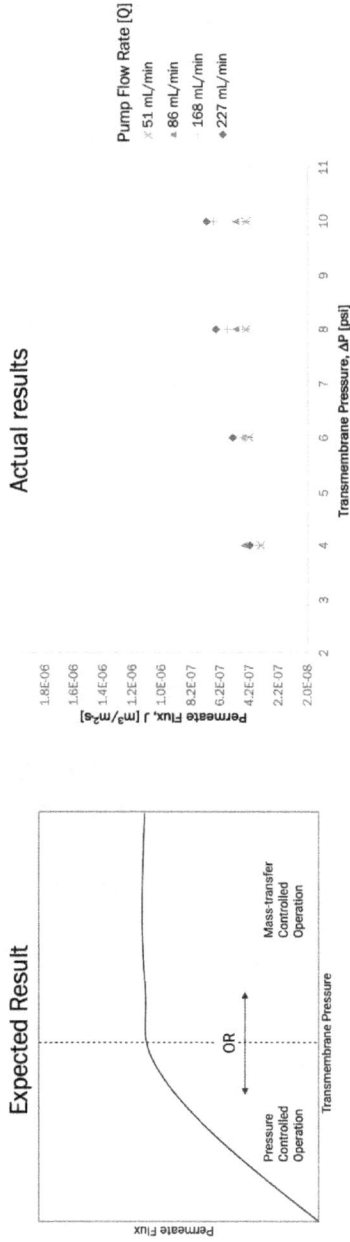

FIGURE 4.1 Example of a memo cover sheet

Example memo body intro. This will begin on its own page following the cover sheet.

Course or Project Name
Agricultural & Biological Engineering
West Lafayette, IN 47907

To: Dr. Goodenough
From: A. B. Estudent
Date: Month Day, Year

Subject: Membrane Control Regime (note that the subject should be the same as the title on the cover sheet)

Membrane filtration is commonly used to purify or concentrate materials. In this case, a pressure gradient was imposed across a semipermeable membrane. The permeate flux was measured as a function of the transmembrane pressure to determine the control regime of the membrane (i.e., if the membrane was pressure or mass-transfer controlled).

The retention coefficient was measured to determine if . . .

(The body of the memorandum continues for an additional one to two pages. Appendices are placed after the conclusion.)

5

APPENDICES

Organizing Supplemental Materials for Clarity and Completeness

---

OBJECTIVES

- Organize supplemental materials to support the main report.
- Differentiate between essential and supplemental content in technical reports.
- Format and label appendices for clarity and accessibility.
- Include sample calculations and raw data appropriately.
- Use nomenclature sections to define symbols and units.

Long derivations and/or other background material that is not essential to the body of the report should be placed in an appendix. The essentials are placed in the body, and the reader is referred to the appendix for details. Tables of data, calibration curves, and supporting graphs are usually placed in an appendix. Sample calculations, which illustrate how the raw data were used to obtain the results, are placed in the appendix to show your reader a well thought-out example of how the calculations were performed. Sample calculations are given more depth at the end of this chapter. The appendices are given descriptive headings to separate different types of material. The appendix should NOT include the entirety of the data tables. If the data is exceptionally brilliant, the full raw data will be sought out.

NOMENCLATURE

The nomenclature section defines all symbols used in the report. Symbols should be listed alphabetically, starting with Latin (Arabic) letters, followed by Greek symbols, and then subscripts and superscripts if applicable. Each entry should include the symbol, its definition, and the units used to help ensure clarity and convenience for the reader. Units are generally based on the **International System of Units, commonly known as SI.**

Example:

SYMBOL	DEFINITION	UNITS
A_C	Cross-sectional area	$[m^2]$
A_S	Surface area	$[m^2]$
K_{og}	Overall gas mass transfer coefficient	$[mol/m^2 \cdot s \cdot atm]$
ΔP	Change in pressure	$[Pa]$
Q	Flow rate	$[m^3/s]$
U	Average gas velocity	$[m/s]$
Re	Reynolds number	$[-]$
t_R	Retention time	$[s]$
μ	Viscosity	$[kg/m \cdot s]$
ρ	Density	$[mg/mL]$

WORKS CITED

References to the literature should be cited in the body of the report using author last name and year of publication enclosed in parentheses. In this case, the references should be listed alphabetically in the Literature Cited section. Refer to Chapter 9 for additional guidance on references and citations.

CONTENT

References are used to help your readers understand from where your information originates and allows them to look up the reference if they would like additional information. If you read something and then use this information to help write your report, it should be referenced at the end of the sentence. *You need to use evidence to support assertions—either experimental evidence or a reliable external resource.* You need to put what you read into your own thoughts. Using more than a few words from a source is considered plagiarism and should be avoided at all costs. The Purdue Online Writing Lab (OWL) is an excellent resource regarding what is plagiarism and how to avoid it. The OWL can help you with citations and references.

You are expected to use author attributed publications for your citations. *You may not use the prelabs, presentations, or lab manuals as a source in your reports.*

1. Only scientific journals, reference books, textbooks, official government publications (e.g., Food and Drug Administration, Centers for Disease Control and Prevention, National Institutes of Health, Environmental Protection Agency, etc.) should be used as references for technical content in your reports.

2. Do NOT use websites such as eHow, Wikipedia, and EngineeringToolbox as your source. There is no guarantee that information on these sites is correct. Some of it can be wrong or misleading. You would never want to reference these types of sites in any sort of professional output. As a student, incorrect or misleading information can significantly undermine your learning.

FORMATTING

In-text citations should be formatted as follows:

Liquid-liquid extraction is used to remove a component of interest from a mixture by employing a second liquid phase (Geankoplis, 2010, p. 776). A scalable method that effectively uses water as an extraction solvent is pressurized liquid extraction (PLE), also known as subcritical fluid extraction. PLE operates with liquids at elevated temperatures, below their critical point, enhancing the extraction kinetics of the solvent and maintaining the solvent in liquid form through increased pressures (Wang & Weller, 2006). Under pressure and elevated temperatures, water becomes a more effective solvent for extracting organic compounds. As water's temperature rises but remains below its critical point, its dielectric constant decreases, reducing its polarity (Ong et al., 2006). Organic compounds are more soluble in less polar solvents, making subcritical pressurized water an optimal extraction solvent for many natural products (Shotipruk et al., 2004).

The resulting references in the literature cited would be formatted as follows:

Geankoplis, C. (2010). *Transport processes and separation process principles*. Prentice Hall.

Ong, E. S., Cheong, J. S. H., & Goh, D. (2006). Pressurized hot water extraction of bioactive or marker compounds in botanicals and medicinal plant materials. *Journal of Chromatography A, 1112*(1–2): 92–102.

Shotipruk, A., Kiatsongserm, J., Pavasant, P., Goto, M., & Sasaki, M. (2004). Pressurized hot water extraction of anthraquinones from the roots of *Morinda citrifolia*. *Biotechnology Progress, 20*(6): 1872–1875.

Wang, L. J., & Weller, C. L. (2006). Recent advances in extraction of nutraceuticals from plants. *Trends in Food Science & Technology, 17*(6): 300–312.

SAMPLE CALCULATIONS

Sample calculations must be included to demonstrate how specific values were derived during your data analysis. These calculations serve two key purposes:

1. They show your instructor or reader how raw data was transformed into meaningful results.
2. They help you, the author, verify your process and avoid propagating errors.

Performing calculations by hand encourages careful attention to unit cancellation and logical flow, reducing the risk of mistakes that can occur when relying solely on spreadsheet formulas. If you input a formula directly into Excel and accept the output without verifying it, you increase the likelihood of error due to hidden assumptions or misapplied logic.

Sample calculations should be handwritten on engineering problems paper or plain paper and included in the appendix of your report. They must be completed in a methodical and easy-to-follow manner. Use units consistently and follow the SKUBASES method (Figure 5.1).

Sample calculations are not considered figures and should not be placed in the main body of the report.

An example of the expected format for sample calculations is provided on the following pages (Figure 5.2). It is important to indicate when a specific final value in your work corresponds to a calculation included in the appendix.

Sketch
Draw the system or dynamics

Known
What is known about the problem? Define variables. Can you suss out relevant information from the problem statement?

Unknown
What is unknown? What are you solving for? What are the variables?

Basic equations
What concepts or standard equations can you use? How does this relate to problems you've already seen in class?

Assumptions
What assumptions are you making to simplify the problem? How can you limit boundaries to make it solvable?

Solution
Solve the problem

$$x = 34.78 \text{ ft}$$

Encircle the answer

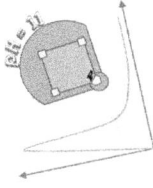

Sanity check
Does the answer make sense? Is it in the right ball park?

PURDUE UNIVERSITY

Agricultural and Biological Engineering

FIGURE 5.1 Common problem-solving method to follow for engineering calculations, adopted for wide use in Purdue University's Department of Agricultural and Biological Engineering

1/3

An outline of the calculations used to obtain final results from the experimental data

Annotations are typed

<u>SAMPLE CALCULATIONS</u>

ALL FOR $Q_{AIR} = 70\, L/min$

Identify the basis for calculation

<u>NH_3 Concentration of Liquid Stream, exit</u>

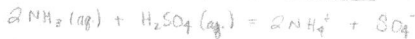

$$2NH_3\,(aq) + H_2SO_4\,(aq.) = 2NH_4^+ + SO_4^=$$

8.5 mL of 0.1 M H_2SO_4 was used to titrate a 25 mL sample

$$C_{NH_3} = \left(\frac{8.5\ mL\ H_2SO_4}{25\ mL\ sample}\right)\left(\frac{0.1\ M\ H_2SO_4}{L}\right)\left(\frac{2\ mol\ NH_3}{mol\ H_2SO_4}\right)$$

$$\boxed{C_{NH_3} = 0.068\ \frac{mol\ NH_3}{L}}$$

Identify what is being calculated

<u>Partial Pressure (P^*) @ Equilibrium w/ exit Liquid</u>

@ 20°C, the Henry's Law Constant $H = 7.37 \times 10^{-3}\,ATM\left(\frac{100\,g\,H_2O}{g\,NH_3}\right)$

State necessary constants and assumptions

$$P_{NH_3} = H C_{NH_3}$$

$$P^* = 7.37 \times 10^{-3}\,ATM\left(\frac{100\,g\,H_2O}{g\,NH_3}\right)\left(\frac{0.068\ mol\ NH_3}{L\cdot solution}\right)\left(\frac{17\,g\,NH_3}{mol\ NH_3}\right)\left(\frac{1\,L}{1000\,g}\right)$$

Show all units

$$\boxed{P^* = 8.55 \times 10^{-4}\ ATM}$$ → the density of the solution ≈ density of H_2O

$$\Delta P_2 = P_{g_2} - P_2^*\ ,\ \text{driving force @ bottom of column}$$

$$P_{g_2} = \left(\frac{2}{102}\right) * 1\,ATM = 0.0196\ ATM$$

$$\boxed{\Delta P_2 = 0.0196\ ATM - 8.55 \times 10^{-4} = 0.0188\ ATM}$$

<u>NH_3 Partial pressure in exit gas</u>

Volume of Bleed gas titrated $= \dfrac{0.380\,L}{min} \cdot \dfrac{1\,min}{60s} \cdot 292s = 1.85\,L$

moles of Bleed gas ⇒ $n = \dfrac{PV}{RT} = \dfrac{1\,ATM \cdot 1.85\,L}{0.0821\frac{L\cdot ATM}{mol\cdot k}(293k)} = 7.70 \times 10^{-2}\,mol$

$n_{NH_3} = $ moles of NH_3 in Bleed gas sample $= \left(\dfrac{0.01\,M\,H_2SO_4}{L}\right)(0.01\,L)\left(\dfrac{2\,mol\,NH_3}{mol\,H_2SO_4}\right)$

$$= 2 \times 10^{-4}\ mol\ NH_3$$

$$\boxed{P_1 = \dfrac{n_{NH_3}}{n_{BleedGas}} \times 1\,ATM = \dfrac{2\times10^{-4}}{7.70\times10^{-2}} \times 1 = 2.60 \times 10^{-3}\ ATM}$$

FIGURE 5.2 Example of proper sample calculations for a report

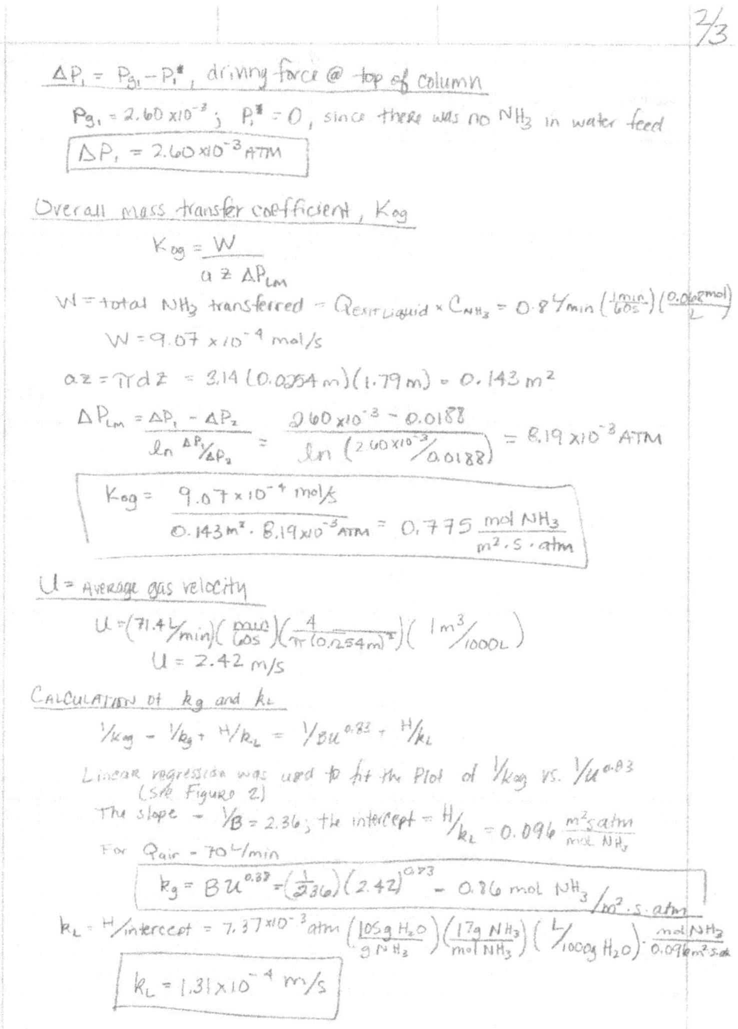

$\Delta P_1 = P_{g_1} - P_1^*$, driving force @ top of column

$P_{g_1} = 2.60 \times 10^{-3}$; $P_1^* = 0$, since there was no NH_3 in water feed

$\boxed{\Delta P_1 = 2.60 \times 10^{-3} \text{ ATM}}$

Overall mass transfer coefficient, K_{og}

$$K_{og} = \frac{W}{a z \, \Delta P_{LM}}$$

$W = $ total NH_3 transferred $= Q_{exit\,liquid} \times C_{NH_3} = 0.87 /min \left(\frac{1 min}{60 s}\right)\left(\frac{0.062 mol}{L}\right)$

$W = 9.07 \times 10^{-4}$ mol/s

$a z = \pi r d \, z = 3.14 (0.0254\,m)(1.79\,m) = 0.143\,m^2$

$\Delta P_{LM} = \frac{\Delta P_1 - \Delta P_2}{\ln \frac{\Delta P_1}{\Delta P_2}} = \frac{2.60 \times 10^{-3} - 0.0188}{\ln \left(2.60 \times 10^{-3} / 0.0188\right)} = 8.19 \times 10^{-3} \text{ ATM}$

$\boxed{K_{og} = \frac{9.07 \times 10^{-4} \text{ mol/s}}{0.143\,m^2 \cdot 8.19 \times 10^{-3} \text{ATM}} = 0.775 \frac{mol\,NH_3}{m^2 \cdot s \cdot atm}}$

$U = $ Average gas velocity

$U = \left(71.4 \,L/min\right)\left(\frac{min}{60 s}\right)\left(\frac{4}{\pi (0.0254\,m)^2}\right)\left(\frac{1\,m^3}{1000\,L}\right)$

$U = 2.42$ m/s

CALCULATION of k_g and k_L

$\frac{1}{k_{og}} = \frac{1}{k_g} + \frac{H}{k_L} = \frac{1}{B U^{0.83}} + \frac{H}{k_L}$

Linear regression was used to fit the plot of $\frac{1}{k_{og}}$ vs. $\frac{1}{u^{0.83}}$ (see Figure 2)

The slope $= \frac{1}{B} = 2.36$; the intercept $= \frac{H}{k_L} = 0.096 \frac{m^2 \cdot atm}{mol\,NH_3}$

For $Q_{air} = 70\,L/min$

$\boxed{k_g = B U^{0.83} = \left(\frac{1}{2.36}\right)(2.42)^{0.83} = 0.86\,mol\,NH_3/m^2 \cdot s \cdot atm}$

$k_L = H/intercept = 7.37 \times 10^{-3} atm \left(\frac{105 g\,H_2O}{g\,NH_3}\right)\left(\frac{17 g\,NH_3}{mol\,NH_3}\right)\left(\frac{1}{1000 g\,H_2O}\right) \frac{mol\,NH_3}{0.096\,m^2 \cdot s \cdot atm}$

$\boxed{k_L = 1.31 \times 10^{-4}\,m/s}$

FIGURE 5.2 *(continued)*

3/3

CALCULATION of k_g using Sherwood-Gilliland Correlation

$$SH = 0.023 \, Re^{0.83} \, Sc^{0.44}$$

$$Sc = \frac{\mu}{\rho_{air} \, D_{NH_3-Air}} = \frac{1.7 \times 10^{-5}}{1.30 \, (2.2 \times 10^{-5})} = 0.595$$

@ $0°C$, $D_{NH_3-Air} = 1.98 \times 10^{-5} \, m^2/s$ $\left(\begin{array}{l}\text{TREYBAL, R.E., Mass Transfer}\\ \text{Operations, } 3^{rd} \text{ Ed. McGraw-Hill,}\\ 1980\end{array}\right)$

$D \propto T^{3/2}$ so @ $20°C$, $D_{NH_3-Air} \left(\frac{293}{273}\right)^{3/2} \times 1.98 \times 10^{-5}$

$$= 2.20 \times 10^{-5} \, m^2/s$$

$\mu_{air} = 0.017 \, C_p = 1.7 \times 10^{-5} \, Pa \cdot s$

$\rho = 1.30 \, kg/m^3$

$$Re = \frac{d u \rho}{\mu} = \frac{(0.0254 m)(2.42 \, m/s)(1.30 \, kg/m^3)}{1.7 \times 10^{-5} \, Pa \cdot s} = 4630$$

$$SH = 0.023 (4630)^{0.83} (0.595)^{0.44} = 20.3 = \frac{k_g \, R T \, d}{D_{NH_3-Air}}$$

$$k_g = \frac{SH \, D_{NH_3-Air}}{R T d} = \frac{20.3 \, (2.2 \times 10^{-5} \, m^2/s)(1.013 \times 10^5 \, Pa/atm)}{(8.314 \frac{m^3 Pa}{mol \cdot K})(293 k)(0.0254 m)}$$

$$\boxed{k_g = 0.073 \frac{mol \, NH_3}{m^2 \cdot s \cdot atm}}$$

Clearly box final answers
or significant calculations

FIGURE 5.2 (*continued*)

6

VISUALS AND DATA DISPLAY

Designing Figures, Tables, and Equations for Technical Reports

OBJECTIVES

- Design and format figures, tables, and equations for technical reports.
- Apply best practices for labeling and captioning visuals.
- Present data clearly using appropriate graphical formats.
- Evaluate and improve the effectiveness of visual communication.
- Critique and revise visuals to improve clarity and impact.

FIGURES

The caption for a figure should be centered below the figure. Captions are placed below a figure because the figure is usually read from the bottom up (i.e., from the axes upward to the trends). The caption should be descriptive enough that the reader can understand the figure without referring to the text. All figures should be referenced in the text before they appear in your report. An example, Figure 6.1, is included in this section. Notice that in the sentence the first letter of the word "Figure" is capitalized when referring to a figure. The reason that Figure 6.1 in the text is capitalized is because now that it has a name it is a proper noun, and proper nouns are capitalized. The font for the figure caption should be 1 point smaller than the font for the report text to differentiate the caption

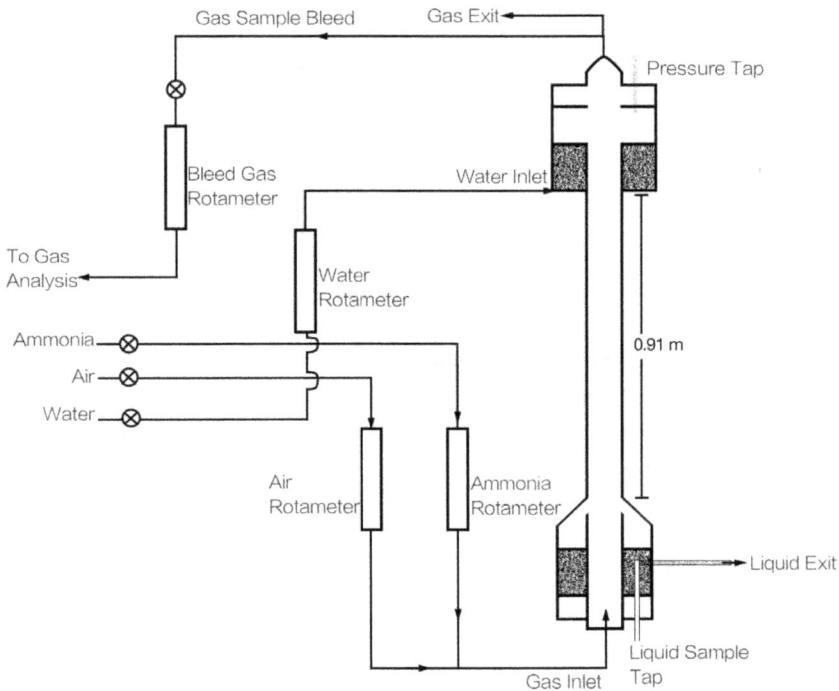

FIGURE 6.1 Schematic diagram of the experimental apparatus to measure the mass transfer coefficient for ammonia absorption

from the report body. Notice the stylization of the caption in Figure 6.1. When referring to your figure, avoid phrases such as "as shown below" and "as seen on the following page" because your figure may not always be placed as intended. It is clear to simply number the figure and refer to it by its number in the text.

Schematics are especially important because they provide a simplified, focused representation of the equipment used in the experiment. Unlike photographs, which mainly show where a piece of equipment is located in a physical space, a schematic highlights the essential components and flow directions and how parts fit together. Dimensions should be included to indicate the size of the equipment, especially when those measurements are used in calculations or are part of the experimental results. A well-drawn schematic demonstrates your understanding of the apparatus and its function within the experiment. Think of the Washington, D.C., metro map, which is a stylized version of the transit system that clearly shows how routes are connected. The map gives the viewer just enough detail to understand how to get from point A to point B. If the viewer had to interpret a full city map with all the streets overlaid, it would be much harder to figure out how to use the system.

ILLUSTRATIONS

1. Do not copy illustrations from any source. *You must make your own schematics.*
   a. Software suggestions to help construct schematics: MS Visio, MS PowerPoint, lucidchart.com, and BioRender.com.
2. Give the drawing a clear title and a figure number. Center the title below the drawing.
3. Label the parts for easy reference. Use arrows if necessary.
4. Depending on the complexity of the drawing, assign numbers or letters to each part with an accompanying key or legend.
5. Include dimensions when necessary.

GRAPHS AND DATA PRESENTATION

Presenting your data in an accurate and concise manner is required for a quality lab report. It is up to you to figure out the best way to display the data analysis from your lab. Figures 6.2 and 6.3 are included as examples of how to present your analysis data. Figure 6.2 is a representation of a inadequate presentation. Figure 6.3 displays a better way to present data. Both graphs were made in Microsoft Excel. The following list shows why Figure 6.2 is **not** the best way to present the data.

- A chart title is given at the top of the graph. This is unnecessary, since your reader can clearly see that the graph shows the wall shear stress as a function of the wall shear rate.
- The units are not included on the axis title.
- The differences in the data series are shown by color difference only. This will not work if your report is printed in black and white.
- The legend takes up a lot of space on the right-hand side.
- The gridlines are distracting.

- The caption leaves a lot to be desired. The figure caption should be descriptive and should allow your figure to stand alone from the text. Your figure caption should convey the following information:
  - What does the graph show?
  - Why is it important?
  - What conclusion can be drawn from the graph?
- The figure caption begins with "Plot of . . . ," which is **unnecessary** since it is generally clear that the figure is a plot.
- The inclusion of regression equations is distracting; they are small, improperly placed, and contain too many significant digits. The importance of fitting the data to an equation is not given in the descriptive caption.
- The acronym (LBG) is not defined.

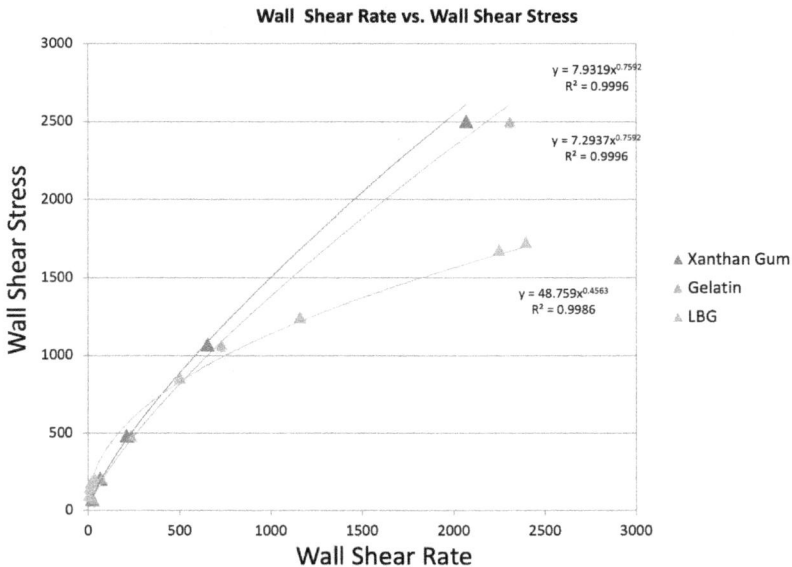

Wall Shear Rate vs. Wall Shear Stress

$y = 7.9319x^{0.7592}$
$R^2 = 0.9996$

$y = 7.2937x^{0.7592}$
$R^2 = 0.9996$

$y = 48.759x^{0.4563}$
$R^2 = 0.9986$

▲ Xanthan Gum
▲ Gelatin
▲ LBG

FIGURE 6.2 Plot of wall shear rate as a function of shear stress. The darkest triangles show the xanthan gum, the medium-dark triangles show the gelatin, and the lightest triangles show the LBG. The data fit well to show shear thinning behavior, and the equations show the consistency index and the flow behavior index.

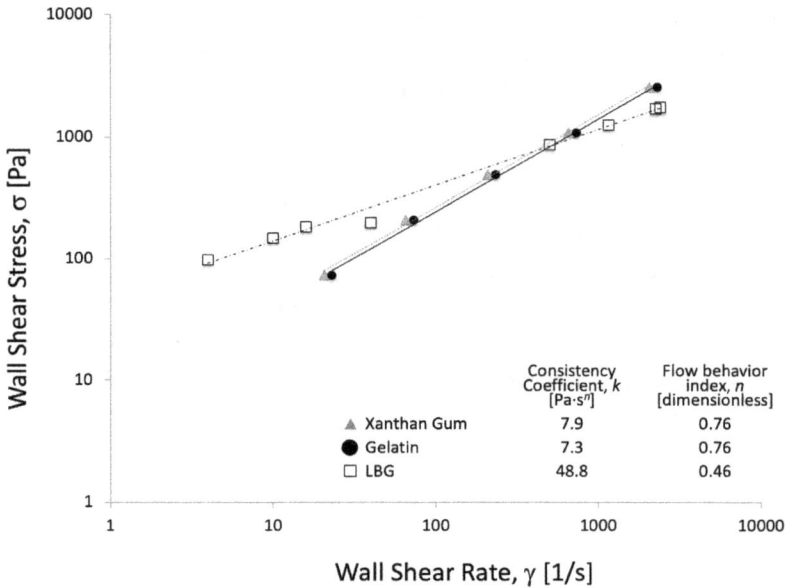

FIGURE 6.3 Behavior profile for xanthan gum, gelatin, and locust bean gum (LBG) determined using a capillary flow viscometer at room temperature (22°C). The profile was plotted as a log-log plot to demonstrate the linear relationship between shear rate and shear stress. Linear regression was performed to determine the fluid type; all regression lines had an R^2 value >0.98 indicating a good fit. All three solutions exhibited shear thinning behavior and was verified using the value of the flow behavior index, n, which was <1 in each case.

TABLES

The caption for a table should be placed above the table and should line up with the left side of the table. The caption is placed above the table, since the table is usually read from the top down (i.e., from the column headers down through the rows). The caption should be descriptive enough that the reader can understand the table without referring to the text. All tables should be referenced in the text before they appear in your report. An example is given in Table 6.1. In text the term "Table" is capitalized when it is followed by a table because now that it has a name, it is a proper noun, and proper nouns are capitalized. The font for the table caption should be 1 point smaller than the font for the report text. Notice the stylization of the caption in Table 6.1. Tables should be well organized and should clearly show the independent variable.

TABLE 6.1. *Experimental results for* K_{og}

Q [L/MIN]	U [M/S]	L [L/MIN]	K_{og} [MOL NH$_3$/M^2· S · ATM]
70	2.42	0.80	0.78*
90	3.12	0.80	1.00
110	3.81	0.80	1.09
130	4.41	0.80	1.21
150	5.19	0.80	1.37

* *A sample calculation for this result is in the sample calculation section of the appendix.*

Special note: Take note of the significant figures (sig figs) in your reports. Many of your calculations will be performed and reported using Microsoft Excel, which will gladly spit out as many sig figs as you want. However, this is not good practice, especially when reporting on experimental data. If the tool you use in the lab is only able to measure one sig fig, then you can only report one sig fig in your analysis. If the tool you used reported three sig figs, then you can use three sig figs.

EQUATIONS

If you add an equation to your report, you **must** refer to it in the text, much like you find in a textbook. You must number your equations. An example is shown in Equation 6.1. Notice that in the previous sentence the first letter of the word "Equation" is capitalized when referring to an equation by number. Now that the equation has a name it is a proper noun, and proper nouns are capitalized. The equation should be tabbed to start at 2 inches from the left margin, and the designation number should be right justified as shown in Equation 6.1. Do not forget to explain the terms in your equations.

$$K_{og} = \frac{W}{a \cdot z \cdot \Delta P_{LM}} \tag{6.1}$$

7

CONVERTING PROCEDURES INTO METHODS SECTIONS

Translating Step-by-Step Instructions into Concise Scientific Descriptions

---

OBJECTIVES

- Distinguish between procedural instructions and method descriptions.
- Translate step-by-step procedures into concise, professional methods sections.
- Identify the level of detail appropriate for technical audiences.
- Use visual aids to support method descriptions.
- Recognize and eliminate unnecessary detail in method descriptions.

Turning a step-by-step procedure into a concise, professional methods section can be surprisingly challenging. While procedures are written to guide someone through an experiment, methods sections are written to document what was done in a clear, efficient manner that avoids unnecessary detail. The goal is to communicate the experimental approach in a way that is reproducible but not instructional. This chapter will help you recognize the differences between procedural and methodological writing and guide you in refining your own work. In this chapter you will find an example procedure, followed by both an overly detailed methods section and a much more concise methods section to illustrate key principles.

EXAMPLE BIOSEPARATIONS LAB PROCEDURE

Here is an example of a written procedure that would be followed during an experiment.

PERISTALTIC PUMP FLOW RATE CALIBRATION

The flow rate of the pump at different settings will need to be determined. It is conventional to list operations, as the flow rate was set to 1.2 mL/min rather than saying the pump was set to 3. You will also need to know the pumping flow rate (m^3/s) for your analysis. The flow rate on the peristaltic pump is adjusted using the knob on the front of the pump drive.

1. Obtain a graduated cylinder, a container of water, and a stopwatch.
2. Use a free piece of flexible tubing to measure your flow rate. Do not flow through the entire system to measure the pumping flow rate.
3. Place the water at the pump inlet and the graduated cylinder at the outlet.
4. Turn the pump on by adjusting the knob.
5. Obtain the flow rate by either recording the time it takes to reach a specified volume or by setting the time and recording the volume. Either way, the time and volume will need to be recorded.
6. Test flow rates between settings 3 and 7. The flow rates between the integer settings can be interpolated from a graph of pump setting (x-axis) versus flow rate (y-axis).
7. It is wise to test each setting at least three times to obtain reliable results.

XANTHAN CONCENTRATION

The concentration of xanthan produced during fermentation can be measured using an oven-drying method, with a viscometer, or with a spectrophotometer. This lab will explore the use of an acid digestion followed by quantification using a spectrophotometer. A stock solution of 0.1 g/L is provided.

CALIBRATION CURVE

1. Retrieve 6 tubes and perform a serial dilution of the stock solution. Label each tube *clearly*.
   a. Serial Dilution Instructions
      i. Tube 1: Add 2 mL of the stock solution.
      ii. Tube 2: Combine 0.75 mL of the stock solution with 0.25 mL of water.
      iii. Tube 3: Remove 1 mL from tube 1 and add to tube 3. Add 1 mL of water. Mix well.
      iv. Tube 4: Remove 1 mL from tube 3 and add to tube 4. Add 1 mL of water. Mix well.
      v. Tube 5: Remove 1 mL from tube 4 and add to tube 5. Add 1 mL of water. Mix well.
      vi. Tube 6: Add 1 mL of water. This will act as your blank or zero concentration tube.
2. Calculate the concentration of the mixture in each tube. IN THE FUME HOOD add 3 mL of concentrated sulfuric acid to each of the tubes containing the xanthan solution. Mix vigorously for 30 seconds. The reaction will increase the temperature of the tubes within 10–15 seconds (Albalasmeh et al., 2013).
3. Cool each tube in an ice bath for 2 minutes to bring to room temperature.
4. Carefully pour the contents of each tube into a cuvette and read in the spectrophotometer at 325 nm.
5. Make a plot of xanthan concentration (x-axis) versus absorbance (y-axis). The points should form a line, and the equation can be used to determine the unknown concentrations in the fermentation samples.
   a. Your lab notebook paper has gridlines; you can easily hand-draw your calibration graph to ensure that your data appears linear.

NOTE: One half of the group can calibrate the spectrophotometer while the other half calibrates the pump.

FILTRATION SAMPLES

1. You will need to perform a dilution to ensure that samples will be within range on the spectrophotometer. Choose a sample you think will have a high xanthan concentration and perform a 20x dilution. Test it on the spectrophotometer and adjust the dilution factor as necessary to make sure your reading

is in the range of your calibration curve. When you have found a dilution factor that works, proceed to dilute and test the remainder of your samples.

2. IN THE FUME HOOD—wearing a lab coat, gloves, and safety glasses— add 3 mL of concentrated sulfuric acid to the tube. Mix vigorously for 30 seconds.

3. Cool in ice for 2 minutes to bring back to room temperature.

4. Carefully pour the digested contents into a cuvette.

5. Read in the spectrophotometer at 325 nm.

6. Use your calibration curve to determine the concentration of xanthan in each sample (*Hint: Remember that you have diluted your samples*). If your absorbance is not within range, increase the dilution factor of your original sample run the procedure again.

FILTRATION OPERATION

An ultrafiltration membrane will be used to concentrate the xanthan gum. Figure 7.1 shows the process for the experiment.

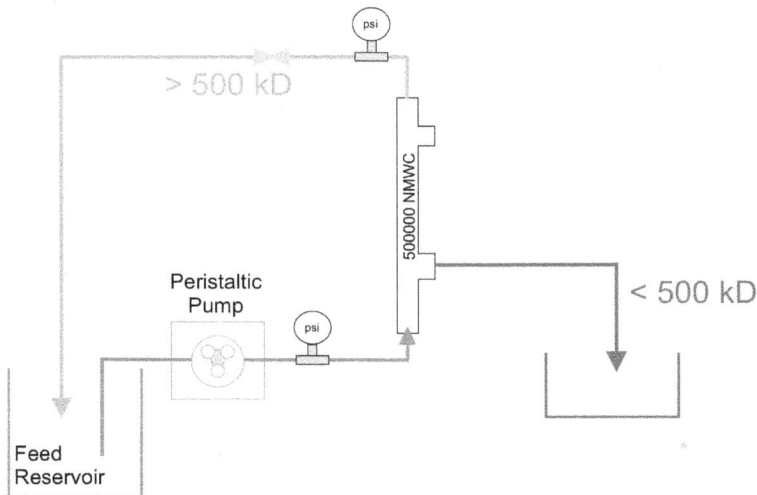

FIGURE 7.1 Schematic of membrane filtration process to concentrate phage from fermentation broth

MEMBRANE OPERATION

For this operation you will vary both the flow rate and the transmembrane pressure.

1. Remove the free tubing from the pump and place the membrane tubing back into the pump head.
2. Obtain your sample and place at the pump inlet. The retentate reservoir outlet will be the same as the feed container. The reservoir should be placed on a heated stir plate set to 30°C.
3. Turn the pump on to the lowest setting for your calibration.
4. Set the potentiometer to induce a transmembrane pressure of 2 psi. Wait until there is liquid flowing from each outlet into the specified reservoir before taking measurements.
5. Obtain a sample from each outlet and reserve for future analysis in the spectrophotometer.
6. Determine the flow rate of the permeate using a graduated cylinder. Use a stopwatch and a graduated cylinder the same as how you measured flow-rate for the pump calibration. You only need to do this once.
7. You will need to repeat steps 4, 5, and 6 at transmembrane pressures of 4, 6, 8, and 10 psi.
8. **Before moving on to the next step**, reduce the transmembrane pressure to 1 psi and run the pump at the 5 setting for 3 minutes to flush the filter.
9. Change the pump flow and repeat the experiment until you have tested four flow rates at five transmembrane pressures. At each new setting, collect samples from the permeate and the retentate and measure the permeate flow rate. This should result in 2 sets of 20 samples.

NOTE: Take an initial concentration reading of your feed. Record the membrane surface area.

EXAMPLE TOO DETAILED METHODS

In this work a 500,000 NMWC ultrafiltration membrane was used to purify 0.5% w/v xanthan solution to determine the selectivity of the membrane and develop an empirical equation relating permeate flux to pump flow rate, transmembrane pressure, and xanthan concentration. The xanthan solution was pumped from a reservoir through a plastic tubing system and into the ultrafiltration membrane

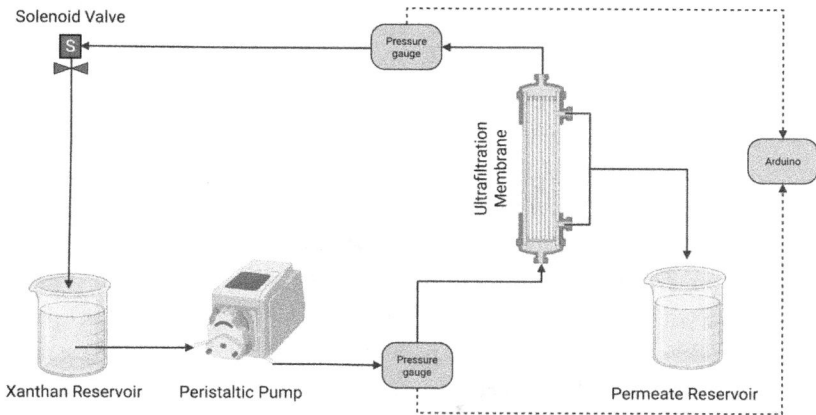

FIGURE 7.2 The process of filtering xanthan solution through an ultrafiltration membrane. Xanthan was pumped through the ultrafiltration membrane and retentate flowed back into the xanthan reservoir. The permeate was separated during pressure variation testing but combined back into the xanthan reservoir before changing flow rate. The Arduino was used to monitor and alter pressure. Created by the author in BioRender (https://BioRender.com).

using a peristaltic pump. The permeate was deposited in a separate container, while the retentate was directed back into the original xanthan reservoir. Transmembrane pressure was measured by taking the average values of two pressure gauges attached to the membrane inlet tube and the retentate return tube as shown in Figure 7.2.

The peristaltic pump flow rate settings were in dimensionless numbers and therefore needed to be related to volumetric flow rate for data analysis. Pump settings 4, 6, 8, and 10 were tested to determine volumetric flow rate by using a stopwatch to measure the time required for the pump output to fill a graduated cylinder to the 5 mL mark with water. Three samples were collected at each pump setting, and the average of these values was used to increase accuracy. Calculations showing flow rate conversions can be found in Appendix C. A calibration curve (Appendix C) was created by graphing average flow rate against pump setting, and a linear trendline was fitted for the interpolation of subsequent pump settings. Transmembrane pressure was controlled using an Arduino microcontroller system. The Arduino controlled pressure by varying the intervals at which the solenoid valve opened and closed, determining how much pressure was released from the system. Desired pressure was set on the Arduino through a potentiometer. The two pressure gauges gave input that was displayed on an LCD screen along with the current pressure setting so the user could adjust the pressure and observe lag time. Testing for different pressure and pump settings was then conducted. Starting at a pump flow rate of 1.61×10^{-8} m³/s and a transmembrane pressure of 2 psi, retentate and permeate samples were collected in microcentrifuge tubes with a volume of 1.5 mL. Additional samples were collected for transmembrane pressures of 4, 6, 8, and 10 psi. The permeate flow rate was then measured using the same method as described to measure the pump flow rate. After all samples at this flow rate were collected, the ultrafiltration membrane was flushed with water at a flow rate of 6.51×10^{-9} m³/s and a pressure of 1 psi for 5 minutes. The permeate collected was combined back with the retentate reservoir so each trial would have a constant xanthan concentration. This procedure was repeated for pump flow rates of 2.259×10^{-9}, 2.902×10^{-9}, and 3.545×10^{-9} m³/s. The 1.5 mL retentate and permeate samples collected in the microcentrifuge tubes were then evaluated for xanthan concentration by spectroscopy. A calibration plot relating absorbance values to concentration of xanthan is shown in Appendix D. The plot was generated with 6 samples of known concentration made through serial dilutions of a 1 g/L stock xanthan solution. The xanthan was added to laboratory tubules followed by 3 mL of sulfuric acid. The tubules were placed on ice for 5 minutes and then poured into individual cuvettes. Absorbance was measured with a spectrophotometer at 325 nm, using water as blank solution before each reading. The absorbance of each collected sample was taken using the same method. However, a 20x dilution was performed on each collected sample prior to testing to ensure that the absorbance value was valid. Xanthan concentration was determined by interpolation from the calibration plot made with stock solution.

EXAMPLE APPROPRIATE METHODS

A 500,000 NMWC ultrafiltration membrane was used to concentrate a 0.5% w/v xanthan solution. Membrane selectivity was evaluated by examining the relationship between permeate flux, pump flow rate, transmembrane pressure, and xanthan concentration. A peristaltic pump (MasterFlex Console Drive) circulated the xanthan solution through the membrane (GE Xampler, Model CFP-2-E-4A, Surface area = 420 cm²) and back into the reservoir, with transmembrane pressure (ΔP) controlled by pressure gauges and a solenoid valve connected to an Arduino unit. Figure 7.3 shows a schematic of the membrane apparatus.

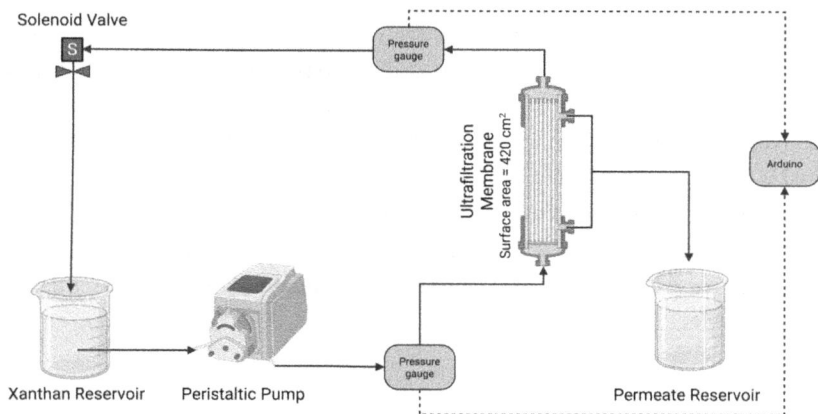

FIGURE 7.3 Schematic diagram of the membrane filtration unit. A 0.5% w/v xanthan solution was concentrated through a 500,000 NMWC ultrafiltration membrane. Transmembrane pressure was controlled through a solenoid valve with input from an Arduino. Pump flow rate was varied through a peristaltic pump. Created by the author in BioRender (https://BioRender.com).

Permeate flux was measured at five pump flow rates (60–120 mL/min) and five transmembrane pressures (ΔP = 2, 4, 6, 8, 10). The peristaltic pump, equipped with an interchangeable head (MasterFlex, Model 07516–12), was calibrated by varying the pump setting and recording the flow rate. The membrane was flushed with DI water for 5 minutes between trials. Xanthan concentration in permeate and retentate was determined using a sulfuric acid hydrolysis method (Albalasmeh et al., 2013). Samples were diluted 20x, hydrolyzed with 3 mL of sulfuric acid for 5 minutes, quenched in an ice bath, and analyzed via spectrophotometer at 325 nm. Calibration curves for pump flow rate and xanthan concentration were used to determine the unknown concentrations in the samples. Figure 2A in the appendix shows the calibration curve relating xanthan concentration to absorbance.

WRAPPING UP

The **appropriate methods section** is more effective because it communicates the experimental approach clearly and concisely, focusing on what was done rather than how to do it. Unlike the overly detailed version, which reads like a procedural manual, the improved version removes unnecessary step-by-step instructions and subjective commentary. It emphasizes reproducibility without overwhelming the reader with minutiae and uses precise language, proper units, and references to equipment and techniques that are relevant to the study's objectives. This style is more aligned with professional scientific writing, where the goal is to document—not instruct—and where clarity and brevity are essential. Figure 7.4 serves as a visual aid to guide you in converting the lab procedure into a well-written methods section. Keep in mind that the procedure may evolve during the experiment, which is why your lab notebook records are essential for accurately detailing what you actually did.

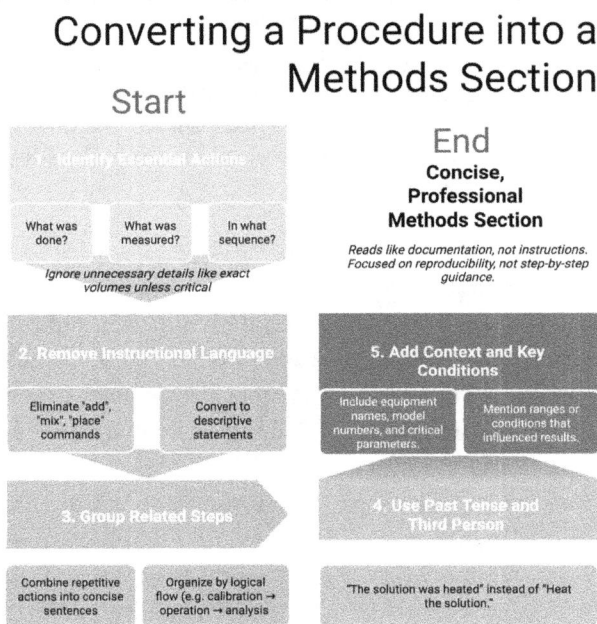

FIGURE 7.4 Basic process to convert a laboratory procedure into an appropriate methods section. Created by the author in BioRender (https://BioRender.com)

REFERENCE USED IN THIS CHAPTER

Albalasmeh, A. A., Berhe, A. A., & Ghezzehei, T. A. (2013). A new method for rapid determination of carbohydrate and total carbon concentrations using UV spectrophotometry. *Carbohydrate polymers, 97*(2), 253–261.

8
EXAMPLES OF TECHNICAL INTRODUCTIONS
Learning from Student Writing and Instructor Feedback

OBJECTIVES

- Analyze examples of effective and ineffective technical introductions.
- Write concise, objective-driven introductions for lab reports.
- Connect experimental goals to broader engineering principles.
- Prepare introductions that align with conclusions.
- Reflect on feedback to improve clarity and alignment in scientific writing.

This chapter presents a series of revised student-written introductions from fermentation lab reports. These examples are included to illustrate the level of writing expected from college juniors and to provide guidance on how to improve scientific communication. Insertions and deletions are shown to highlight how original student work can be refined for clarity, precision, and professionalism. Footnotes accompany each introduction to offer insight into the types of feedback instructors may provide during grading or revision. The final example is an exemplary introduction that students can aspire to emulate, although it is not intended as a definitive model. Footnotes for the exemplar explain why it is effective, helping students understand the qualities of strong scientific writing.

EXAMPLE INTRODUCTION A

Temperature is an important factor <u>in microbial fermentation, influencing both cell growth and product formation.</u>[1] ~~When fermenting D-glucose into butanol, temperature? can be used to test for the growth rate.~~ <u>This experiment investigates the effect of temperature on the fermentation of D-glucose by</u> *Clostridium acetobutylicum* <u>to produce n-butanol.</u>[2] The goals of the fermentation experiment were to ~~ascertain the temperature effects—at both 25°C and 30°C—on the growth of Clostridium acetobutylicum and the production of butanol from D-glucose and discover the yield of the experiment as a function of temperature~~ <u>evaluate how two temperatures—25°C and 30°C—affect cell growth and butanol yield over a 26-hour fermentation period].</u>[3] ~~To do this, three phase cultures were inoculated at separate times and used in the experiment. The phase 3 culture was inoculated 26 hours before the first lab, phase 2 culture was inoculated 13 hours before the first lab and phase 1 culture was inoculated in the first lab with the inoculation time set at 0.~~ <u>Three cultures were inoculated at staggered intervals (0, 13, and 26 hours prior to the experiment) to capture a full growth curve].</u>[4] ~~Two distinct temperatures for reaction occurrence 25°C and 30°C were observed.~~ <u>Fermentation was conducted at both temperatures, and samples were collected every 45 minutes].</u>[5] Samples were ~~taken every 45 minutes and were clustered around time zero, 13 hours, and 26 hours to observe the full growth curve~~ <u>grouped around 0 h, 13 h, and 26 h to monitor fermentation progress].</u>[6] The relative cell density, D-glucose concentration, and butanol concentration were determined for each sample.[7] ~~The experimental process and data analysis employed to derive relationships pertaining to the goals of the experiment follow.~~ <u>These measurements were used to calculate specific growth rates and product yields, enabling a comparative analysis of temperature-dependent fermentation kinetics].</u>[8]

1. Be specific as to why temperature is important to the experiment in question.
2. Clarifies the biological context and removes ambiguity about the role of temperature.
3. This is a very long sentence. Be more succinct with your goals. This is a short experiment and report, and your objective should be concise.
4. I don't understand what you mean by this. Phase cultures? It appears that you tested a complete time series that was split into three inoculation periods. Why not just say that you tested the effect of temperature on fermentation over a period of ~26 hours?

5. This sentence is unnecessary. At your mention of temperature at any other place in this paragraph you can put the two that were compared.
6. Improves clarity.
7. Add detail regarding methods used to measure the concentrations.
8. Could have a stronger closing statement to the introduction. You want to be sure to write your introduction to be able to tie in with your conclusion.

EXAMPLE INTRODUCTION B

Glucose fermentation is an anaerobic biological process many[1] microbes use in the production of energy[2] (Prescott et al., 2008). Products of fermentation[3] can include substances such as organic acids, alcohols, and polysaccharides. Butanol is the desired product of interest from the bacterial fermentation of ~~In the case of~~ *Clostridium acetobutylicum* ~~fermentation, the product created is butanol~~(Weber et al., 2005). One of the main factors as to how much product is created in a fermentation process is temperature.[4,5] Therefore,[6] the examination of both bacterial growth rate and yield according to varying temperatures is needed when designing industrial butanol production {because the temperature can control how much product is created}.[7] ~~Evaluation of t~~The effect of ~~varying~~ temperature—25°C and 35°C—was ~~done~~ evaluated by measuring glucose concentration, butanol concentration, and cell density ~~by analyzing *C. acetobytylicum* growth at 25°C and 35°C~~ over the span of ~~around~~ 26 hours. Samples were taken from each culture~~, and continued to be taken~~ every 45 minutes for a total of ~~five~~ 15 samples ~~from each inoculation~~ in groups of five clustered around 0 h, 13 h, and 26 hr. Every sample was then evaluated for glucose concentration, butanol concentration, and cell density.[8] Substrate and product concentration and cell density were used to ~~This information was then used to~~ develop relationships comparing xanthan gum concentration, specific growth rate ~~of the bacterial cells, and~~ and butanol yield ~~between the two~~ at each temperature~~s over time~~.

1. Many is a generic term. Can you be more explicit or precise? consortia? community?
2. How does it produce energy? Fermentation uses energy to convert some type of food into an output.
3. But you just stated that the product of fermentation was energy, and now you are saying that the products are organic substances. Please resolve.

4. What are some other factors that are important? Substrate? Agitation?
5. (citation needed)
6. Why should this sentence begin with "therefore"? This does not logically flow.
7. This is redundant with the previous sentence.
8. Add detail regarding methods used to measure the concentrations.

EXAMPLE INTRODUCTION C

Fermentation is a biological process in which a high energy compound[1] is consumed in the absence of oxygen to produce chemical compounds such as organic acids, gases, or alcohols.[2] This process has been widely used to preserve food like kimchi and yogurt, as well as to produce alcoholic drinks like wine and beer.[2] Butanol is the major product of interest formed ~~by~~ during the fermentation of glucose by *Clostridium acetobutylicum* (Jones & Woods, 1986). Butanol is used as a fuel and solvent for an ~~wide spread~~ array of industrial applications.[2] Therefore,[3] there is a great desire to optimize the production of butanol (Qureshi & Blaschek, 2006). Butanol is produced as a by-product of fermentation.[4] ~~The process of fermentation may lead some cells to lyse (Weiss & Ollis, 1979).~~[5] The yield and production rate of butanol ~~significantly~~ depend[6] on temperature ~~according to what has been seen in literature~~[7] (Lee et al., 2008). Higher[8] temperature can result in higher yield of butanol, while lower temperature increases cell growth of the bacterium (Lee et al., 2008). Hence,[9] kinetic parameters of fermentation such as specific cell growth coefficient and changes in concentration of butanol will be studied in at ~~small lab scale~~ shake-flask scale ~~in order~~ to determine the effect of temperature on butanol production rate and ~~cell growth~~ yield. Understanding the ~~relationship~~ interaction of temperature with growth and ~~stated above will allow~~ yield allows for a strong ~~the~~ prediction of the ~~how much~~ resources required to scale the process to obtain a specific yield. ~~are need to obtain a sufficient yield in upscaled projects.~~

1. What do you mean "high energy." It is usually just a sugar. Why is this considered to be high?
2. Note that "like" means similar to but not including, and "such as" means similar to and including. Also, this statement needs a citation.
3. Are you saying that because its use is widespread it has to be optimized? If it is valuable enough then that may not be the case.

4. This does not make sense. You stated previously that butanol is the major product, but now you are saying it is a by-product? Resolve this for me. Also the correct spelling is by-product.

5. What does cell lysis have to do with what you are talking about?

6. You do not need the word "significantly" here. Be careful when using the word "significant" in a technical context because it has a specified meaning in statistics and data analysis.

7. No need to say "according to literature" if you include your source.

8. Define higher and lower. Do you have a range for this? Can higher be $>30°C$?

9. The sentence does not solicit a "hence." Please rephrase to improve flow.

EXEMPLAR INTRODUCTION

Temperature is a critical parameter in microbial fermentation, influencing both metabolic activity and product yield.[1] The effect of temperature on the production of butanol, a valuable biofuel and industrial solvent, was investigated using *Clostridium acetobutylicum*, an anaerobic bacterium known for its ability to produce acetone, butanol, and ethanol through acetone-butanol-ethanol (ABE) fermentation (Jones & Woods, 1986).[2]

Butanol production is sensitive to environmental conditions, with temperature playing a key role in regulating enzymatic activity, solventogenesis, and cell viability.[3] Lower temperatures (e.g., 25°C) may favor biomass accumulation, while higher temperatures (e.g., 35°C) can enhance solvent production, although excessive heat may inhibit microbial growth (Qureshi & Blaschek, 2006; Lee et al., 2008).[4]

Fermentation was conducted at two temperatures—25°C and 35°C—to evaluate the impact on butanol yield and microbial growth over a 48-hour period.[5] Cultures were sampled at regular intervals to measure cell density, glucose consumption, and butanol concentration.[6] These data were used to calculate specific growth rates and product yields, enabling a comparative analysis of temperature-dependent fermentation kinetics.[7] Understanding the relationship between temperature and butanol production is essential for optimizing bioprocesses and improving the economic feasibility of biofuel manufacturing.[8]

1. Clear topic sentence that introduces the relevance of temperature in microbial fermentation. It sets the stage for the rest of the paragraph.

2. The specific organism and product are introduced early, with proper scientific naming and citation. This grounds the experiment in a known biological context.

3. Mechanistic insight into how temperature affects fermentation, showing depth of understanding.

4. Balanced explanation of expected outcomes at different temperatures, supported by literature.

5. Concise experimental objective stated in third person, with clear variables and time frame.

6. Relevant measurements are listed, showing how data will be collected to support the objective.

7. Connection to analysis—explains how the data will be used to draw conclusions.

8. Strong closing statement that ties the experiment to broader industrial relevance, reinforcing the importance of the study.

WRAPPING UP

The examples given in this chapter are to help you see what technical writing looks like at different stages of development. By comparing revised student work with an exemplar, you can begin to understand what clarity, precision, and professionalism look like in practice. The footnotes are meant to show the kinds of feedback you might receive—some structural, some content-specific—and how that feedback is intended to help you improve your writing.

Good technical writing is not innate. It is built through revision and reflection. When you take time to consider feedback and compare your writing to strong examples, you start to notice patterns—what you are doing well and where you can grow. This kind of reflection leads to writing that better supports your experimental goals and communicates your findings more effectively.

One of the best ways to improve is to read more technical writing. Journal articles, textbooks, and manuals all model the tone, structure, and vocabulary used in professional engineering contexts. The more you read, the more familiar these conventions become and the easier it is to apply them in your own work.

Ultimately, learning to write well in technical settings takes practice. These examples, the feedback, and your own reflections are tools to help you build confidence and competence as a communicator.

9
REFERENCES AND CITATIONS
Using Sources Responsibly and Professionally in Engineering Writing

OBJECTIVES

- Understand the importance of citing sources in technical writing.
- Differentiate between citations and references.
- Apply appropriate citation styles for engineering documents.
- Use reference management tools to organize sources.
- Evaluate source credibility and avoid inappropriate references.

WHY CITE YOUR SOURCES?

Citing sources is crucial because it demonstrates that you have conducted thorough research, lends credibility and support to your findings and recommendations, allows others to reproduce your research process, and acknowledges the contributions of others. Proper citation is essential for backing up assertions made in your reports with evidence from external resources or experimental findings. All statements of fact should be supported either using experimental data or valid external resources.

CITATION VERSUS REFERENCE

A citation informs readers where the information originated, while a reference provides detailed information about the source, enabling readers to understand the type of source and locate it if necessary. Citations are included in the text, whereas references are typically listed at the end of the document.

WHAT IS ALLOWED?

Strive to only use author-attributed publications for citations. Acceptable sources include scientific journals, reference books, textbooks, and official government publications (e.g., Food and Drug Administration, Centers for Disease Control and Prevention, National Institutes of Health, Environmental Protection Agency). These sources ensure the technical content in your reports is reliable and credible.

WHAT IS NOT ALLOWED?

Do not use prelabs, presentations, or lab manuals as sources in your reports. Additionally, websites such as eHow, Wikipedia, and EngineeringToolbox are not acceptable sources due to the lack of guaranteed accuracy.

WHERE TO FIND GOOD SOURCES?

Good sources can be found using databases such as Web of Science, Google Scholar, and your institution's library resources. These platforms provide access to a wide range of credible peer-reviewed scientific literature and are essential for building a strong foundation for your work.

In addition to traditional databases, tools powered by artificial intelligence (AI) are becoming increasingly valuable for literature discovery and reference management, tools such as

- Scite, which helps evaluate how a paper has been cited (supporting, contradicting, or mentioning), making it easier to assess credibility and relevance;

- Research Rabbit, which allows you to visually explore networks of related research papers and authors, helping you uncover connections and trends; and
- Litmaps, which offers interactive maps of research topics and citation networks, making it easier to track the development of ideas and identify key papers.

These tools can help you quickly find reputable sources, identify gaps in the literature, and stay organized as you build your reference list. They are especially useful when starting a new project or exploring unfamiliar topics.

Note: AI tools for literature discovery are evolving rapidly. New features, platforms, and integrations are emerging all the time. Stay curious and keep exploring. What is available today may look very different in just a few months. Always verify the credibility of sources and cross-check with established databases when using AI-assisted tools.

REFERENCE MANAGERS

Managing citations and references can quickly become overwhelming, especially in multisource research projects. Fortunately, several tools are available to help you stay organized and format your references correctly:

- Mendeley (free). Offers citation management, PDF annotation, and collaboration features.
- Zotero (free). A powerful open-source tool that automatically extracts citation data from web pages and integrates with Word and Google Docs.
- EndNote (paid). A robust tool often used in professional and academic settings, with advanced features for managing large libraries.
- Microsoft Word (free with Office). Includes basic citation and bibliography tools that are useful for shorter or simpler projects.
- EasyBib (free). A web-based tool that helps generate citations in various formats, ideal for quick reference tasks.

These tools help you store, organize, and format references efficiently, saving time and reducing errors in citation style. Most integrate directly with word processors, allowing you to insert citations and generate bibliographies with just a few clicks.

Note on AI Integration: Many reference managers and literature discovery platforms are now incorporating AI-powered features to enhance search, organization, and citation accuracy. For example, tools such as Scite, Research Rabbit, and Litmaps use AI to help you find relevant reputable sources and visualize connections between papers. These technologies are evolving rapidly, so staying up to date with new features and platforms can give you a significant edge in your research workflow.

10

LABORATORY NOTEBOOK GUIDELINES

Documenting Experiments for Reproducibility and Professional Practice

---

OBJECTIVES

- Recognize the role of lab notebooks in engineering documentation and IP protection.
- Maintain detailed, legible, and reproducible records of experiments.
- Apply best practices for organizing and formatting notebook entries.
- Use structured rubrics to assess and improve notebook quality.

RATIONALE

Whether in industry, academic, or government work environments, laboratory notebooks are critical documents for research and engineering design. They are *primary legal documentation of inventions* and can be the basis for supporting intellectual property and patent applications. They are also the documentation that allows anyone (including you at a later date) to repeat and perform the same procedures in the future. Science and engineering are founded in this principle, called *autonomous replication*. Our entire framework of science and engineering is based on only accepted scientific results that can be reliably replicated by others.

FEATURES OF A GOOD LAB NOTEBOOK

- Detailed enough that another person of comparable skill could repeat your experiment and obtain the same results using only the lab notebook.
- Written in blue or black *ink*.
  - ○ Do not erase or scribble out incorrect information. Simply ~~strikethrough~~ with a single line and initial the change.
- Contains enough detail that you can effectively troubleshoot your procedures should you obtain unexpected results.
- Is neat and easy to read by you and others.
- Makes use of tables, figures, photographs, and drawings to organize data and illustrate procedures, setup, observations, and results.
- Each page contains a date and a place for your signature at the bottom.
- Uses headers to organize all required sections.
  - ○ Aim/purpose
  - ○ Materials and Equipment Setup
    - ◆ Includes process diagram or drawing of equipment setup and a comprehensive list of estimated amount or volume of chemicals and solvents.
  - ○ Procedures and Protocols
    - ◆ Includes all variables and order that they were varied.
    - ◆ Should be detailed and step-by-step accounting of what you plan to do.
    - ◆ Include in-lab changes to protocol. There may be times when you will need to deviate from the given manual.
  - ○ Results
    - ◆ All raw data
    - ◆ Include your leading zeros when recording results. Excluding the leading zero provides an easy opportunity for later data falsification.
    - ◆ Organize numbers into tables when appropriate.
    - ◆ Make and document observations.
    - ◆ Sketch/graph data as you go.
    - ◆ Tape in all photographs, images, printouts, etc. Date and initial each taped-in item.
  - ○ Calculations that were made for the procedures
    - ◆ For example, dilution calculations, mass to mole conversions for weigh outs, etc.
- Analysis and Interpretation

- ○ Document your thoughts and observations on the results.
- ○ Include your rationale and your thought process as to why you did what you did.
- ○ Include what you think the data meant at the time you are recording it.
- ○ Include preliminary conclusions from the experiment.

LAB NOTEBOOK OVERVIEW

A-Level Work

Contains all *Features of a Good Lab Notebook*. Contains signatures, extremely neat; complete protocols; thorough observations; all data and calculations; and all images and drawings such that someone could repeat the work from documentation.

B-Level Work

Written with needed headers and may be missing some detail but contains most of the data. However, someone might have difficulty repeating the experiment.

C-Level Work

Log style only (no headers; e.g., objectives, rationale, results, next steps) and may contain data but overall is insufficient to repeat the experiment.

D-Level Work

Extremely brief, missing most major components.
Clear that content was recorded well after the actual experiment.
Impossible to repeat the experiment as presented.

F-Level Work

Entire experimental sections or entire lab notebook missing.

An example lab notebook checklist is contained herein.

NOTEBOOK CHECK

STUDENT:

NOTEBOOK ITEM	NOT PRESENT	POOR	FAIR	GOOD	EXCELLENT
Contains Experimental Aim(s)					
Contains Materials and Setup (where applicable)					
• *Process diagram or equipment setup where applicable* • *List of solvents and amounts*					
Contains Procedures and Protocols					
• *All variables and order varied* • *Detailed, step by step* • *Includes in-lab changes*					
Contains Results • *Raw data* • *Leading zeros* • *Tables* • *Observations* • *Graphs*					
Contains Calculations (where applicable)					
• *Calculations made for procedures (in lab)*					
Contains Analysis • *Thoughts on results* • *Preliminary conclusions* • *Rationale/ thought process*					

Overall Housekeeping • *Neat and easy to read* • *Pages signed and dated* • *Single strike-through mistakes, initialed and dated* • *Legible* • *Written in pen* • *Detailed*					

Notes:

TA Signature: Date:

FINAL REPORT SUBMISSION CHECKLIST

Use this checklist to review most any report before submission. If you can confidently check each box, your report is likely in great shape.

Formatting and Structure

- Report uses correct font, line spacing, and margins.
- Section headings follow the required format (Summary, Introduction, Methods, etc.).
- Each major section begins on a new page where required.
- Figures and tables are numbered and referenced in the text.

Content and Clarity

- **Beginning**: Clearly sets the stage with goals, context, and relevant background. Explains why the work matters and what the reader should expect.
- **Middle**: Describes what was done in a clear, organized way. Includes essential details about the experiment, equipment, and data collected without unnecessary procedural steps.
- **End**: Interprets the data, presents results clearly, and ties findings back to the original goals. Includes thoughtful discussion of limitations and a conclusion that summarizes key takeaways.

Visuals and Data

- All figures and tables are **labeled, captioned,** and **easy to interpret.**
- Axes include **units** and **descriptive titles**.
- Visuals are referenced in the text before they appear.
- Significant figures are appropriate for the precision of the instruments used.

Technical Writing Quality

- Writing is **concise, clear,** and **objective**.
- Report is written in **third person** and **past tense**.
- No contractions (e.g., use "do not" instead of "don't").
- Acronyms are spelled out at first use.
- All variables and uncommon terms are defined.

References and Attribution

- All external sources are **cited properly** in the text.
- A **Literature Cited** section is included and formatted consistently.
- No direct quotes—information is paraphrased in your own words.
- Only **credible sources** are used (e.g., journals, textbooks, government publications).

Appendices and Calculations

- Sample calculations are included in the appendix and follow the SKUBASES format.
- Raw data, calibration curves, and supplemental visuals are organized and labeled.
- Nomenclature section defines all symbols and units used in the report.

Professional Presentation

- Report is **neat, organized,** and **easy to read**.
- No spelling or grammar errors.
- All pages are numbered and include your name and submission date.

ADDITIONAL RESOURCES

Much of my own education on technical writing has stemmed from *The ACS Style Guide: Effective Communication of Scientific Information* by Anne M. Coghill and Lorrin R. Garson.

The technical writing process described in Chapter 1 was inspired by *Technical Writing Process: The Simple, Five-Step Guide That Anyone Can Use to Create Technical Documents Such as User Guides, Manuals, and Procedures* by Kieran Morgan.

FIGURE ACKNOWLEDGMENT

Figures 1.1, 7.2, 7.3, and 7.4 were created in BioRender by Abigail S. Engelberth, 2025, https://BioRender.com.

ABOUT THE AUTHOR

ABIGAIL S. ENGELBERTH is an associate professor in the Department of Agricultural and Biological Engineering at Purdue University. Her work focuses on sustainable bioprocessing and developing innovative ways to transform agricultural and food wastes into higher-value products. She is passionate about mentoring students and creating hands-on learning experiences that spark curiosity and creativity. Engelberth enjoys designing projects that connect classroom concepts to real-world challenges, helping students see the impact of engineering on sustainability.

www.ingramcontent.com/pod-product-compliance
Lightning Source LLC
Chambersburg PA
CBHW071119210326
41519CB00020B/6350